Fundamentals of OPTICS

Fundamentals of
OPTICS

Dr. E. J. Holmyard

Oxford Book Company
Jaipur ▪ New Delhi

Published By

Oxford Book Company

Regd. Off. "Bony Residency", Gate No. 2
Opp. Tilak Public School, Vishweswariya Nagar,
Gopalpura Road, Jaipur - 302018, India
Ph: 0141 - 2761280, Telefax: 2761381
email - oxfordbookjpr@gmail.com
website - oxfordbookcompany.com

Delhi Office: 102, 1st Floor, Satyam House,
4327/3, Ansari Road, Darya Ganj,
New Delhi - 110002, India
Ph: 011-45652440

Distributed by
KEYA INTERNATIONAL PRESS LLP
E-mail: keyainternationalpressllp@gmail.com

Published in India, 2026

ISBN : 9789355241528

Typeset by Shivangi Computers, Jaipur

Printed and bound in India by Replika Press Pvt. Ltd.

Contents

Preface

At the dawn of the 20th century, few scientists, even in their wildest dreams, could have imagined the momentous revolution ahead. The discovery of atoms and particicles, the notion of space time, the equivalence of matter and energy, the wave nature of particles and the particle nature of light, new quantum phases of matter, semiconductors, superconductors, the expanding universe... All of these are part of the great scientific revolution of the past 100 years, a revolution that continues today, and whose consequences are rapidly transforming our world and impacting your life. Most optical phenomena can be accounted for using the classical electromagnetic description of light. Complete electromagnetic descriptions of light are, however, often difficult to apply in practice. Practical optics is usually done using simplified models. The most common of these, geometric optics, treats light as a collection of rays that travel in straight lines and bend when they pass through or reflect from surfaces. Physical optics is a more comprehensive model of light, which includes wave effects such as diffraction and interference that cannot be accounted for in geometric optics. Historically, the ray-based model of light was developed first, followed by the wave model of light. Progress in electromagnetic theory in the 19th century led to the discovery that light waves were in fact electromagnetic radiation. Some phenomena depend on the fact that light has both wave-like and particle-like properties. Explanation of these effects requires quantum mechanics. When considering light's particle-like properties, the light is modelled as a collection of particles called "photons". Quantum optics deals with the application of quantum mechanics to optical systems. Optical science is relevant to and studied in many related disciplines including astronomy, various engineering fields, photography, and medicine (particularly ophthalmology and optometry). Practical applications of optics are found in a variety of technologies and everyday objects, including mirrors, lenses, telescopes, microscopes, lasers, and fibre optics.

Reflections can be divided into two types: specular reflection and diffuse reflection. Specular reflection describes the gloss of surfaces such as mirrors, which reflect light in a simple, predictable way. This allows for production of reflected images that can be associated with an actual (real) or extrapolated (virtual) location in space. Diffuse reflection describes opaque, non limpid materials, such as paper or rock. The reflections from these surfaces can only be described statistically, with the exact distribution of the reflected light depending on the microscopic structure of the material. Many diffuse reflectors are described or can be approximated by Lambert's cosine law, which describes surfaces that have equal luminance when viewed from any angle. Glossy surfaces can give both specular and diffuse reflection. In specular reflection, the direction of the reflected ray is determined by the angle the incident ray makes with the surface normal, a line perpendicular to the surface at the point where the ray hits. The incident and reflected rays and the normal lie in a single plane, and the angle between the reflected ray and the surface normal is the same as that between the incident ray and the normal. This is known as the Law of Reflection. For flat mirrors, the law of reflection implies that images of objects are upright and the same distance behind the mirror as the objects are in front of the mirror. The image size is the same as the object size. The law also implies that mirror images are parity inverted, which we perceive as a left-right inversion. Images formed from reflection in two (or any even number of) mirrors are not parity inverted. Corner reflectors retroreflect light, producing reflected rays that travel back in the direction from which the incident rays came. Mirrors with curved surfaces can be modelled by ray-tracing and using the law of reflection at each point on the surface. For mirrors with parabolic surfaces, parallel rays incident on the mirror produce reflected rays that converge at a common focus. Other curved surfaces may also focus light, but with aberrations due to the diverging shape causing the focus to be smeared out in space. In particular, spherical mirrors exhibit spherical aberration. Curved mirrors can form images with magnification greater than or less than one, and the magnification can be negative, indicating that the image is inverted. An upright image formed by reflection in a mirror is always virtual, while an inverted image is real and can be projected onto a screen.

The present book has been designed to outline the basic and fundamental aspects of this subject to be understood in its right perspective. The book uses straight forward, less-technical jargon and manages to introduce each chapter with a basic concept, which ultimately evolves into a more specific detailed principle.

— ***Dr. E.J. Holmyard***

1

Foundational Optics

Modelling and Design of Optical Systems using Physical Optics

Many simplified approximations are available for analysing and designing optical systems. Most of these use a single scalar quantity to represent the electric field of the light wave, rather than using a vector model with orthogonal electric and magnetic vectors. The Huygens–Fresnel equation is one such model. This was derived empirically by Fresnel in 1815, based on Huygen's hypothesis that each point on a wavefront generates a secondary spherical wavefront, which Fresnel combined with the principle of superposition of waves. The Kirchoff diffraction equation, which is derived using Maxwell's equations, puts the Huygens-Fresnel equation on a firmer physical foundation. Examples of the application of Huygens–Fresnel principle can be found in the sections on diffraction and Fraunhofer diffraction.

More rigorous models, involving the modelling of both electric and magnetic fields of the light wave, are required when dealing with the detailed interaction of light with materials where the interaction depends on their electric and magnetic properties. For instance, the behaviour of a light wave interacting with a metal surface is quite different from what happens when it interacts with a di-electric material. A vector model must also be used to model polarised light.

Numerical modelling techniques such as the Finite element method, the Boundary element method and the Transmission-line matrix method can be used to model the propagation of light in systems which cannot be solved analytically. Such models are computationally demanding and are normally only used to solve small-scale problems that require accuracy beyond that which can be achieved with analytical solutions. All of the results from geometrical optics can be recovered using the techniques of Fourier optics which apply many of the same mathematical and analytical techniques used in acoustic engineering and signal processing. Gaussian beam propagation is a simple paraxial physical optics model for the propagation of coherent radiation such as laser beams. This technique partially accounts for diffraction, allowing accurate calculations of the rate at which a laser beam expands with distance, and the minimum size to which the beam can be focused. Gaussian beam propagation thus bridges the gap between geometric and physical optics.

Superposition and Interference

In the absence of nonlinear effects, the superposition principle can be used to predict the shape of interacting waveforms through the simple addition of the disturbances. This interaction of waves to produce a resulting pattern is generally termed "interference" and can result in a variety of outcomes.

If two waves of the same wavelength and frequency are *in phase*, both the wave crests and wave troughs align. This results in constructive interference and an increase in the amplitude of the wave, which for light is associated with a brightening of the waveform in that location. Alternatively, if the two waves of the same wavelength and frequency are out of phase, then the wave crests will align with wave troughs and vice-versa. This results in destructive interference and a decrease in the amplitude of the wave, which for light is associated with a dimming of the waveform at that location.

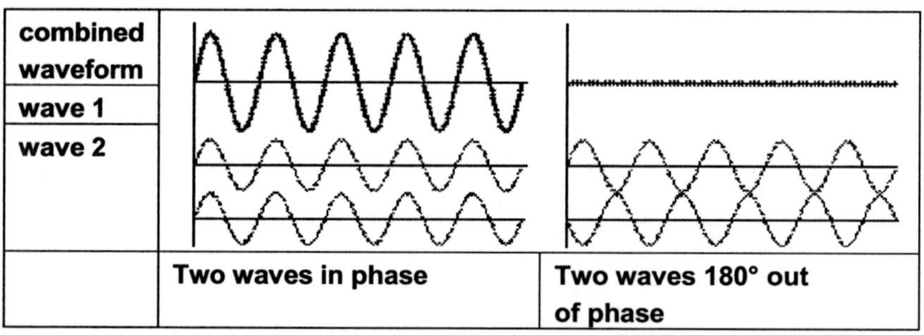

Since the Huygens–Fresnel principle states that every point of a wavefront is associated with the production of a new disturbance, it is possible for a wavefront to interfere with itself constructively or destructively at different locations producing bright and dark fringes in regular and predictable patterns. Interferometry is the science of measuring these patterns, usually as a means of making precise determinations of distances orangular resolutions. The Michelson interferometre was a famous instrument which used interference effects to accurately measure the speed of light.

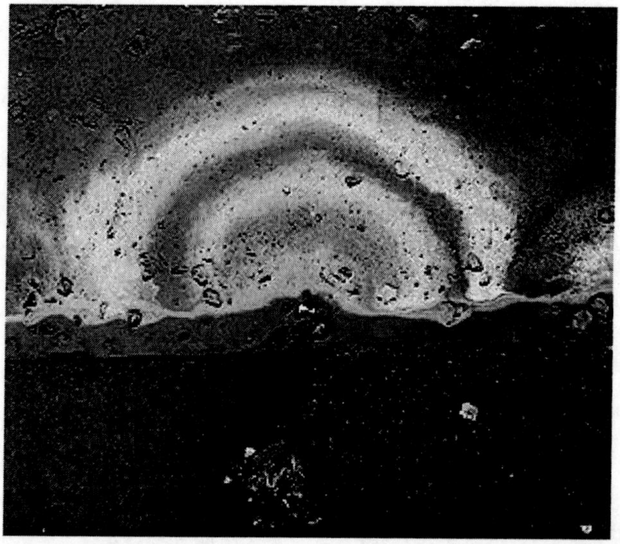

Figure: When oil or fuel is spilled, colourful patterns are formed by thin-film interference.

The appearance of thin films and coatings is directly affected by interference effects. Antireflective coatings use destructive interference to reduce the reflectivity of the surfaces they coat, and can be used to minimise glare and unwanted reflections. The simplest case is a single layer with thickness one-fourth the wavelength of incident light. The reflected wave from the top of the film and the reflected wave from the film/material interface are then exactly 180° out of phase, causing destructive interference. The waves are only exactly out of phase for one wavelength, which would typically be chosen to be near the centre of the visible spectrum, around 550 nm. More complex designs using multiple layers can achieve low reflectivity over a broad band, or extremely low reflectivity at a single wavelength.

Constructive interference in thin films can create strong reflection of light in a range of wavelengths, which can be narrow or broad depending on the design of the coating. These films are used to make dielectric mirrors, interference filters, heat reflectors, and filters for colour separation in colour television cameras. This interference effect is also what causes the colourful rainbow patterns seen in oil slicks.

Diffraction and Optical Resolution

Diffraction is the process by which light interference is most commonly observed. The effect was first described in 1665 by Francesco Maria Grimaldi, who also coined the term from the Latin *diffringere*, 'to break into pieces'. Later that century, Robert Hooke and Isaac Newton also described phenomena now known to be diffraction in Newton's rings while James Gregory recorded his observations of diffraction patterns from bird feathers.

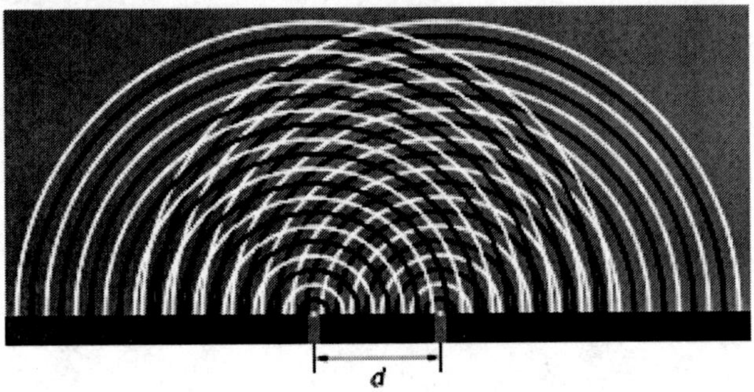

Figure: Diffraction on two slits separated by distance d. The bright fringes occur along lines where black lines intersect with black lines and white lines intersect with white lines. These fringes are separated by angle q and are numbered as order n.

The first physical optics model of diffraction that relied on the Huygens–Fresnel principle was developed in 1803 by Thomas Young in his interference experiments with the interference patterns of two closely spaced slits. Young showed that his results could only be explained if the two slits acted as two unique sources of waves rather than corpuscles. In 1815 and 1818, Augustin-Jean Fresnel firmly established the mathematics of how wave interference can account for diffraction. The simplest physical models of diffraction use equations that describe the angular separation of light and dark fringes due to light of a particular wavelength (λ). In general, the equation takes the form

$$m\lambda = d\sin\theta$$

where d is the separation between two wavefront sources (in the case of Young's experiments, it wastwo slits), θ is the angular separation between the central fringe and the mth order fringe, where the central maximum is $m = 0$.

This equation is modified slightly to take into account a variety of situations such as diffraction through a single gap, diffraction through multiple slits, or diffraction through a diffraction grating that contains a large number of slits at equal spacing. More complicated models of diffraction require working with the mathematics of Fresnel or Fraunhofer diffraction. X-ray diffraction makes use of the fact that atoms in a crystal have regular spacing at distances that are on the order of one angstrom. To see diffraction patterns, x-rays with similar wavelengths to that spacing are passed through the crystal. Since crystals are three-dimensional objects rather than two-dimensional gratings, the associated diffraction pattern varies in two directions according to Bragg reflection, with the associated bright spots occurring in unique patterns and d being twice the spacing between atoms.

Diffraction effects limit the ability for an optical detector to optically resolve separate light sources. In general, light that is passing through an aperture will experience diffraction and the best images that can be created (as described in diffraction-limited optics) appear as a central spot with surrounding bright rings, separated by dark nulls; this pattern is known as an Airy pattern, and the central bright lobe as an Airy disk. The size of such a disk is given by

$$\sin\theta = 1.22\frac{\lambda}{D}$$

where θ is the angular resolution, λ is the wavelength of the light, and D is the diametre of the lens aperture. If the angular separation of the two points is significantly less than the Airy disk angular radius, then the two points cannot be resolved in the image, but if their angular separation is much greater than this, distinct images of the two points are formed and they can therefore be resolved. Rayleigh defined the somewhat arbitrary "Rayleigh criterion" that two points whose angular separation is equal to the Airy disk radius (measured to first null, that is, to the first place where no light is seen) can be considered to be resolved. It can be seen that the greater the diametre of the lens or its aperture, the finer the resolution. Interferometry, with its ability to mimic extremely large baseline apertures, allows for the greatest angular resolution possible.

For astronomical imaging, the atmosphere prevents optimal resolution from being achieved in the visible spectrum due to the atmospheric scattering and dispersion which cause stars to twinkle. Astronomers refer to this effect as the quality of astronomical seeing. Techniques known as adaptive optics have been utilised to eliminate the atmospheric disruption of images and achieve results that approach the diffraction limit.

Dispersion and Scattering

Refractive processes take place in the physical optics limit, where the wavelength of light is similar to other distances, as a kind of scattering. The simplest type of scattering is Thomson scattering which occurs when electromagnetic waves are deflected by single particles. In the limit of Thompson scattering, in which the wavelike nature of light is evident, light is dispersed independent of the frequency, in contrast to Compton scattering which is frequency-dependent and strictly a quantum mechanical process, involving the nature of light as particles.

In a statistical sense, elastic scattering of light by numerous particles much smaller than the wavelength of the light is a process known as Rayleigh scattering while the similar process for scattering by particles that are similar or larger in wavelength is known as Mie scattering with the Tyndall effect being a commonly observed result.

A small proportion of light scattering from atoms or molecules may undergo Raman scattering, wherein the frequency changes due to excitation of the atoms and molecules. Brillouin scattering occurs when the frequency of light changes due to local changes with time and movements of a dense material.

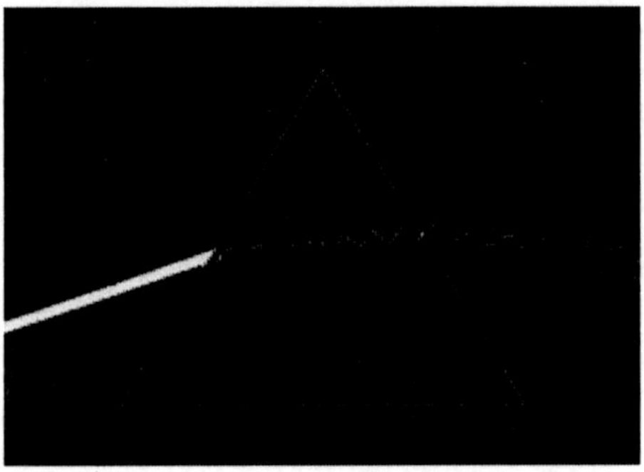

Figure: Conceptual animation of light dispersion through a prism. High frequency (blue) light is deflected the most, and low frequency (red) the least.

Dispersion occurs when different frequencies of light have different phase velocities, due either to material properties (*material dispersion*) or to the geometry of an optical waveguide (*waveguide dispersion*). The most familiar form of dispersion is a decrease in index of refraction with increasing wavelength, which is seen in most transparent materials. This is called "normal dispersion". It occurs in all dielectric materials, in wavelength ranges where the material does not absorb light. In wavelength ranges where a medium has significant absorption, the index of refraction can increase with wavelength. This is called "anomalous dispersion".

The separation of colours by a prism is an example of normal dispersion. At the surfaces of the prism, Snell's law predicts that light incident at an angle θ to the normal will be refracted at an angle arcsin (sin (θ)/ n). Thus, blue light, with its higher refractive index, is bent more strongly than red light, resulting in the well-known rainbow pattern.

Dispersion: Two sinusoids propagating at different speeds make a moving interference pattern. The red dot moves with the phase velocity, and the green dots propagate with the group velocity. In this case, the phase velocity is twice the group velocity. The red dot overtakes two green dots, when moving from the left to the right of the figure. In effect, the individual waves (which travel with the phase velocity) escape from the wave packet (which travels with the group velocity).

Material dispersion is often characterised by the Abbe number, which gives a simple measure of dispersion based on the index of refraction at three specific wavelengths. Waveguide dispersion is dependent on the propagation constant. Both kinds of dispersion cause changes in the group characteristics of the wave, the features of the wave packet that change with the same frequency as the amplitude of the electromagnetic wave. "Group velocity dispersion" manifests as a spreading-out of the signal "envelope" of the radiation and can be quantified with a group dispersion delay parameter:

$$D = \frac{1}{v_g^2}\frac{dv_g}{d\lambda}$$

where v_g is the group velocity. For a uniform medium, the group velocity is

$$v_g = c\left(n - \lambda\frac{dn}{d\lambda}\right)^{-1}$$

where n is the index of refraction and c is the speed of light in a vacuum. This gives a simpler form for the dispersion delay parameter:

$$D = -\frac{\lambda}{c}\frac{d^2 n}{d\lambda^2}.$$

If D is less than zero, the medium is said to have *positive dispersion* or normal dispersion. If D is greater than zero, the medium has *negative dispersion*. If a light pulse is propagated through a normally dispersive medium, the result is the higher frequency components slow down more than the lower frequency components. The pulse therefore becomes *positively chirped*, or *up-chirped*, increasing in frequency with time. This causes the spectrum coming out of a prism to appear with red light the least refracted and blue/violet light the most refracted. Conversely, if a pulse travels through an anomalously (negatively) dispersive medium, high frequency components travel faster than the lower ones, and the pulse becomes *negatively chirped*, or *down-chirped*, decreasing in frequency with time. The result of group velocity dispersion, whether negative or positive, is ultimately temporal spreading of the pulse. This makes dispersion management extremely important in optical communications systems based on optical fibres, since if dispersion is too high, a group of pulses representing information will each spread in time and merge together, making it impossible to extract the signal.

Polarisation

Polarisation is a general property of waves that describes the orientation of their oscillations. Fortransverse waves such as many electromagnetic waves, it describes the orientation of the oscillations in the plane perpendicular to the wave's direction of travel. The oscillations may be oriented in a single direction (linear polarisation), or the oscillation direction may rotate as the wave travels (circular orelliptical polarisation). Circularly polarised waves can rotate rightward or leftward in the direction of travel, and which of those two rotations is present in a wave is called the wave's chirality.

The typical way to consider polarisation is to keep track of the orientation of the electric field vector as the electromagnetic wave propagates. The electric field vector of a plane wave may be arbitrarily divided into two perpendicular components labelled x and y (with z indicating the direction of travel). The shape traced out in the x-y plane by the electric field vector is a Lissajous figure that describes the polarisation *state*. The following figures show some examples of the evolution of the electric field vector (blue), with time (the vertical axes), at a particular point in space, along with its x and y *components* (red/left and green/right), and the path traced by the vector in the plane (purple): The same evolution would occur when looking at the electric field at a particular time while evolving the point in space, along the direction opposite to propagation.

The x and y components of the light wave are in phase. In this case, the ratio of their strengths is constant, so the direction of the electric vector (the vector sum of these two components) is constant. Since the tip of the vector traces out a single line in the plane, this special case is called linear polarisation. The direction of this line depends on the relative amplitudes of the two components.

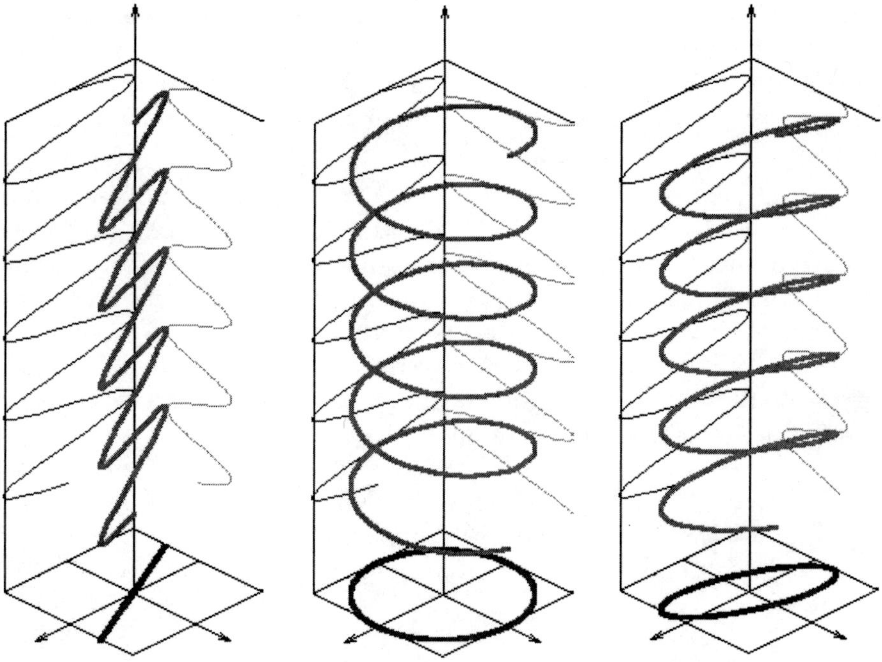

Linear Circular Elliptical polarisation

The two orthogonal components have the same amplitudes and are 90° out of phase. In this case, one component is zero when the other component is at maximum or minimum amplitude. There are two possible phase relationships that satisfy this requirement: the x component can be 90° ahead of the y component or it can be 90° behind the y component. In this special case, the electric vector traces out a circle in the plane, so this polarisation is called circular polarisation. The rotation direction in the circle depends on which of the two phase relationships exists and corresponds to *right-hand circular polarisation* and *left-hand circular polarisation.* In all other cases, where the two components either do not have the same amplitudes and/or their phase difference is neither zero nor a multiple of 90°, the polarisation is called elliptical polarisation because the electric vector traces out an ellipse in the plane (the *polarisation ellipse*). This is shown in the above figure on the right. Detailed mathematics of polarisation is done using Jones calculus and is characterised by the Stokes parameters.

Media that have different indexes of refraction for different polarisation modes are called *birefringent. Well* known manifestations of this effect appear in optical wave plates/retarders (linear modes) and in Faraday rotation/optical rotation (circular modes). If the path length in the birefringent medium is sufficient, plane waves will exit the material with a significantly different propagation direction, due torefraction. For example, this is the case with macroscopic crystals of calcite, which present the viewer with two offset, orthogonally polarised images of whatever is viewed through them.

It was this effect that provided the first discovery of polarisation, by Erasmus Bartholinus in 1669. In addition, the phase shift, and thus the change in polarisation state, is usually frequency dependent, which, in combination with dichroism, often gives rise to bright colours and rainbow-like effects. In mineralogy, such properties, known as pleochroism, are frequently exploited for the purpose of identifying minerals using polarisation microscopes. Additionally, many plastics that are not normally birefringent will become so when subject to mechanical stress, a phenomenon which is the basis of photoelasticity. Non-birefringent methods, to rotate the linear polarisation of light beams, include the use of prismatic polarisation rotators which utilise total internal reflection in a prism set designed for efficient colinear transmission.

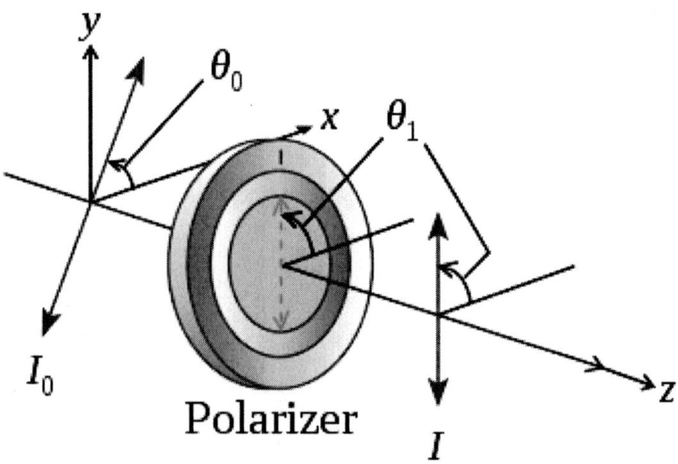

Figure: A polariser changing the orientation of linearly polarised light.

In this picture, $\theta_1 - \theta_0 = \theta_i$.

Media that reduce the amplitude of certain polarisation modes are called dichroic with devices that block nearly all of the radiation in one mode known as *polarising filters* or simply "polarisers". Malus' law, which is named after Etienne-Louis Malus, says that when a perfect polariser is placed in a linear polarised beam of light, the intensity, I, of the light that passes through is given by

$$I = I_0 \cos^2 \theta_i \quad ,$$

where

I_0 is the initial intensity,

and θ_i is the angle between the light's initial polarisation direction and the axis of the polariser. A beam of unpolarised light can be thought of as containing a uniform mixture of linear polarisations at all possible angles. Since the average value of cos $^2\theta$ is 1/2, the transmission coefficient becomes

$$\frac{I}{I_0} = \frac{1}{2}$$

In practice, some light is lost in the polariser and the actual transmission of unpolarised light will be somewhat lower than this, around 38% for Polaroid-type polarisers but considerably higher (>49.9%) for some birefringent prism types.

In addition to birefringence and dichroism in extended media, polarisation effects can also occur at the (reflective) interface between two materials of different refractive index. These effects are treated by the Fresnel equations. Part of the wave is transmitted and part is reflected, with the ratio depending on angle of incidence and the angle of refraction. In this way, physical optics recovers Brewster's angle.

Most sources of electromagnetic radiation contain a large number of atoms or molecules that emit light. The orientation of the electric fields produced by these emitters may not be correlated, in which case the light is said to be unpolarised. If there is partial correlation between the emitters, the light is partially *polarised.* If the polarisation is consistent across the spectrum of the source, partially polarised light can be described as a superposition of a completely unpolarised component, and a completely polarised one. One may then describe the light in terms of the degree of polarisation, and the parameters of the polarisation ellipse.

Light reflected by shiny transparent materials is partly or fully polarised, except when the light is normal (perpendicular) to the surface. It was this effect that allowed the mathematician Etienne Louis Malus to make the measurements that allowed for his development of the first mathematical models for polarised light. Polarisation occurs when light is scattered in the atmosphere. The scattered light produces the brightness and colour in clear skies. This partial polarisation of scattered light can be taken advantage of using polarising filters to darken the sky in photographs. Optical polarisation is principally of importance in chemistry due to circular dichroism and optical rotation ("*circular birefringence*") exhibited by optically active (chiral) molecules.

Modern Optics

Modern optics encompasses the areas of optical science and engineering that became popular in the 20th century. These areas of optical science typically relate to the electromagnetic or quantum properties of light but do include other topics.

A major subfield of modern optics, quantum optics, deals with specifically quantum mechanical properties of light. Quantum optics is not just theoretical; some modern devices, such as lasers, have principles of operation that depend on quantum mechanics. Light detectors, such as photomultipliers and channeltrons, respond to individual photons.

Electronic image sensors, such as CCDs, exhibit shot noise corresponding to the statistics of individual photon events. Light-emitting diodes and photovoltaic cells, too, cannot be understood without quantum mechanics. In the study of these devices, quantum optics often overlaps with quantum electronics.

Speciality areas of optics research include the study of how light interacts with specific materials as in crystal optics and metamaterials. Other research focuses on the phenomenology of electromagnetic waves as in singular optics, non-imaging optics, non-linear optics, statistical optics, and radiometry.

Additionally, computer engineers have taken an interest in integrated optics, machine vision, and photonic computing as possible components of the "next generation" of computers.

Today, the pure science of optics is called optical science or optical physics to distinguish it from applied optical sciences, which are referred to as optical engineering. Prominent subfields of optical engineering include illumination engineering, photonics, and optoelectronics with practical applications like lens design, fabrication and testing of optical components, and image processing. Some of these fields overlap, with nebulous boundaries between the subjects terms that mean slightly different things in different parts of the world and in different areas of industry. A professional community of researchers in nonlinear optics has developed in the last several decades due to advances in laser technology.

Lasers

A laser is a device that emits light (electromagnetic radiation) through a process called *stimulated emission*. The term *laser* is an acronym for *Light Amplification by Stimulated Emission of Radiation*. Laser light is usually spatially coherent, which means that the light either is emitted in a narrow, low-divergence beam, or can be converted into one with the help of optical components such as lenses.

Because the microwave equivalent of the laser, the *maser*, was developed first, devices that emit microwave and radio frequencies are usually called *masers*.

The first working laser was demonstrated on 16 May 1960 by Theodore Maiman at Hughes Research Laboratories. When first invented, they were called "a solution looking for a problem". Since then, lasers have become a multi-billion dollar industry, finding utility in thousands of highly varied applications. The first application of lasers visible in the daily lives of the general population was the supermarket barcode scanner, introduced in 1974.

The laserdisc player, introduced in 1978, was the first successful consumer product to include a laser, but the compact disc player was the first laser-equipped device to become truly common in consumers' homes, beginning in 1982.

These optical storage devices use a semiconductor laser less than a millimetre wide to scan the surface of the disc for data retrieval. Fibre-optic communication relies on lasers to transmit large amounts of information at the speed of light.

Other common applications of lasers include laser printers and laser pointers. Lasers are used in medicine in areas such as bloodless surgery, laser eye surgery, and laser capture microdissection and in military applications such as missile defence systems, electro-optical countermeasures (EOCM), and LIDAR. Lasers are also used in holograms, bubblegrams, laser light shows, and laser hair removal.

Applications

Optics is part of everyday life. The ubiquity of visual systems in biology indicate the central role optics plays as the science of one of the five senses. Many people benefit from eyeglasses or contact lenses, and optics are integral to the functioning of many consumer goods including cameras. Rainbows and mirages are examples of optical phenomena. Optical communication provides the backbone for both the Internet and modern telephony.

Human Eye

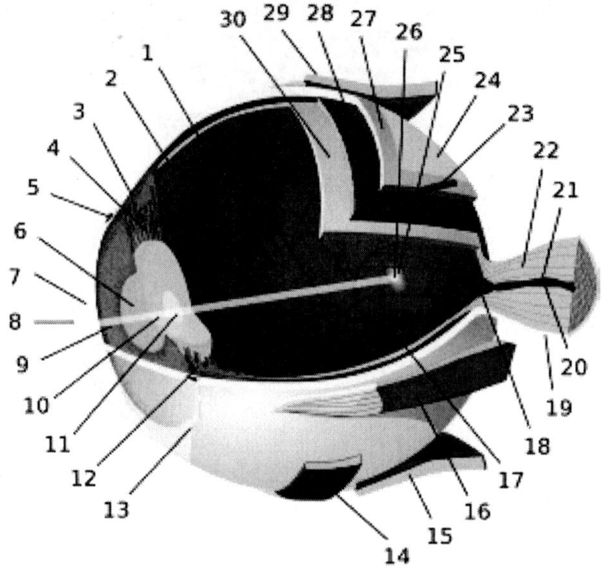

Figure: Model of a human eye. Features mentioned in this article are 3. ciliary muscle, 6. pupil, 8. cornea, 10. lens cortex, 22.optic nerve, 26. fovea, 30. retina

The human eye functions by focusing light onto an array of photoreceptor cells called theretina, which covers the back of the eye. The focusing is accomplished by a series of transparent media. Light entering the eye passes first through the cornea, which provides much of the eye's optical power. The light then continues through the fluid just behind the cornea—the anterior chamber, then passes through the pupil. The light then passes through the lens, which focuses the light further and allows adjustment of focus. The light then passes through the main body of fluid in the eye—the vitreous humour, and reaches the retina. The cells in the retina cover the back of the eye, except for where the optic nerve exits; this results in a blind spot.

There are two types of photoreceptor cells, rods and cones, which are sensitive to different aspects of light. Rod cells are sensitive to the intensity of light over a wide frequency range, thus are responsible for black-and-white vision. Rod cells are not present on the fovea, the area of the retina responsible for central vision, and are not as responsive as cone cells to spatial and temporal changes in light. There are, however, twenty times more rod cells than cone cells in the retina because the rod cells are present across a wider area. Because of their wider distribution, rods are responsible for peripheral vision. In contrast, cone cells are less sensitive to the overall intensity of light, but come in three varieties that are sensitive to different frequency-ranges and thus are used in the perception of colour and photopic vision. Cone cells are highly concentrated in the fovea and have a high visual acuity meaning that they are better at spatial resolution than rod cells. Since cone cells are not as sensitive to dim light as rod cells, most night vision is limited to rod cells.

Likewise, since cone cells are in the fovea, central vision (including the vision needed to do most reading, fine detail work such as sewing, or careful examination of objects) is done by cone cells. Ciliary muscles around the lens allow the eye's focus to be adjusted. This process is known as accommodation.

The near point and far point define the nearest and farthest distances from the eye at which an object can be brought into sharp focus. For a person with normal vision, the far point is located at infinity. The near point's location depends on how much the muscles can increase the curvature of the lens, and how inflexible the lens has become with age. Optometrists, ophthalmologists, and opticians usually consider an appropriate near point to be closer than normal reading distance—approximately 25 cm.

Defects in vision can be explained using optical principles. As people age, the lens becomes less flexible and the near point recedes from the eye, a condition known as presbyopia. Similarly, people suffering from hyperopia cannot decrease the focal length of their lens enough to allow for nearby objects to be imaged on their retina.

Conversely, people who cannot increase the focal length of their lens enough to allow for distant objects to be imaged on the retina suffer from myopia and have a far point that is considerably closer than infinity. A condition known as astigmatism results when the cornea is not spherical but instead is more curved in one direction. This causes horizontally extended objects to be focused on different parts of the retina than vertically extended objects, and results in distorted images.

All of these conditions can be corrected using corrective lenses. For presbyopia and hyperopia, a converging lens provides the extra curvature necessary to bring the near point closer to the eye while for myopia a diverging lens provides the curvature necessary to send the far point to infinity. Astigmatism is corrected with a cylindrical surface lens that curves more strongly in one direction than in another, compensating for the non-uniformity of the cornea. The optical power of corrective lenses is measured in diopters, a value equal to the reciprocal of the focal length measured in metres; with a positive focal length corresponding to a converging lens and a negative focal length corresponding to a diverging lens. For lenses that correct for astigmatism as well, three numbers are given: one for the spherical power, one for the cylindrical power, and one for the angle of orientation of the astigmatism.

Visual Effects

Figure: The Ponzo Illusion relies on the fact that parallel lines appear to converge as they approach infinity.

Optical illusions (also called visual illusions) are characterised by visually perceived images that differ from objective reality. The information gathered by the eye is processed in the brain to give a percept that differs from the object being imaged. Optical illusions can be the result of a variety of phenomena including physical effects that create images that are different from the objects that make them, the physiological effects on the eyes and brain of excessive stimulation (e.g. brightness, tilt, colour, movement), and cognitive illusions where the eye and brain make unconscious inferences.

Cognitive illusions include some which result from the unconscious misapplication of certain optical principles. For example, the Ames room, Hering, Müller-Lyer, Orbison, Ponzo, Sander, and Wundt illusions all rely on the suggestion of the appearance of distance by using converging and diverging lines, in the same way that parallel light rays (or indeed any set of parallel lines) appear to converge at a vanishing point at infinity in two-dimensionally rendered images with artistic perspective. This suggestion is also responsible for the famous moon illusion where the moon, despite having essentially the same angular size, appears much larger near the horizon than it does at zenith. This illusion so confounded Ptolemy that he incorrectly attributed it to atmospheric refraction when he described it in his treatise, *Optics*.

Another type of optical illusion exploits broken patterns to trick the mind into perceiving symmetries or asymmetries that are not present. Examples include the café wall, Ehrenstein, Fraser spiral, Poggendorff, and Zφllner illusions. Related, but not strictly illusions, are patterns that occur due to the superimposition of periodic structures. For example transparent tissues with a grid structure produce shapes known as moiré patterns, while the superimposition of periodic transparent patterns comprising parallel opaque lines or curves produces line moiré patterns.

Optical Instruments

Single lenses have a variety of applications including photographic lenses, corrective lenses, and magnifying glasses while single mirrors are used in parabolic reflectors and rear-view mirrors. Combining a number of mirrors, prisms, and lenses produces compound optical instruments which have practical uses. For example, a periscope is simply two plane mirrors aligned to allow for viewing around obstructions. The most famous compound optical instruments in science are the microscope and thetelescope which were both invented by the Dutch in the late 16th century.

Microscopes were first developed with just two lenses: an objective lens and an eyepiece. The objective lens is essentially a magnifying glass and was designed with a very small focal length while the eyepiece generally has a longer focal length. This has the effect of producing magnified images of close objects. Generally, an additional source of illumination is used since magnified images are dimmer due to the conservation of energy and the spreading of light rays over a larger surface area. Modern microscopes, known as *compound microscopes* have many lenses in them (typically four) to optimize the functionality and enhance image stability. A slightly different variety of microscope, the comparison microscope, looks at side-by-side images to produce a stereoscopic binocular view that appears three dimensional when used by humans.

The first telescopes, called *refracting telescopes* were also developed with a single objective and eyepiece lens. In contrast to the microscope, the objective lens of the telescope was designed with a large focal length to avoid optical aberrations. The objective focuses an image of a distant object at its focal point which is adjusted to be at the focal point of an eyepiece of a much smaller focal length. The main goal of a telescope is not necessarily magnification, but rather collection of light which is determined by the physical size of the objective lens. Thus, telescopes are normally indicated by the diametres of their objectives rather than by the magnification which can be changed by switching eyepieces. Because the magnification of a telescope is equal to the focal length of the objective divided by the focal length of the eyepiece, smaller focal-length eyepieces cause greater magnification.

Since crafting large lenses is much more difficult than crafting large mirrors, most modern telescopes are *reflecting telescopes*, that is, telescopes that use a primary mirror rather than an objective lens. The same general optical considerations apply to reflecting telescopes that applied to refracting telescopes, namely, the larger the primary mirror, the more light collected, and the magnification is still equal to the focal length of the primary mirror divided by the focal length of the eyepiece. Professional telescopes generally do not have eyepieces and instead place an instrument (often a charge-coupled device) at the focal point instead.

Photography

The optics of photography involves bothlenses and the medium in which the electromagnetic radiation is recorded, whether it be a plate, film, or charge-coupled device. Photographers must consider the reciprocity of the camera and the shot which is summarised by the relation Exposure ”” ApertureArea × ExposureTime × SceneLuminance

In other words, the smaller the aperture (giving greater depth of focus), the less light coming in, so the length of time has to be increased (leading to possible blurriness if motion occurs). An example of the use of the law of reciprocity is the Sunny 16 rule which gives a rough estimate for the settings needed to estimate the proper exposure in daylight. A camera's aperture is measured by a unitless number called the f-number or f-stop,*f*/#, often notated as N, and given by

$$f/\# = N = \frac{f}{D}$$

where f is the focal length, and D is the diametre of the entrance pupil. By convention, "*f*/#" is treated as a single symbol, and specific values of *f*/# are written by replacing the number sign with the value. The two ways to increase the f-stop are to either decrease the diametre of the entrance pupil or change to a longer focal length (in the case of a zoom lens, this can be done by simply adjusting the lens). Higher f-numbers also have a larger depth of field due to the lens approaching the limit of a pinhole camera which is able to focus all images perfectly, regardless of distance, but requires very long exposure times.

The field of view that the lens will provide changes with the focal length of the lens. There are three basic classifications based on the relationship to the diagonal size of the film or sensor size of the camera to the focal length of the lens:

- Normal lens: angle of view of about 50° (called *normal* because this angle considered roughly equivalent to human vision) and a focal length approximately equal to the diagonal of the film or sensor.
- Wide-angle lens: angle of view wider than 60° and focal length shorter than a normal lens.
- Long focus lens: angle of view narrow than a normal lens. This is any lens with a focal length longer than the diagonal measure of the film or sensor. The most common type of long focus lens is the telephoto lens, a design that uses a special *telephoto group* to be physically shorter than its focal length.

Modern zoom lenses may have some or all of these attributes.

The absolute value for the exposure time required depends on how sensitive to light the medium being used is (measured by the film speed, or, for digital media, by the quantum efficiency). Early photography used media that had very low light sensitivity, and so exposure times had to be long even for very bright shots. As technology has improved, so has the sensitivity through film cameras and digital cameras.

Other results from physical and geometrical optics apply to camera optics. For example, the maximum resolution capability of a particular camera set-up is determined by the diffraction limit associated with the pupil size and given, roughly, by the Rayleigh criterion.

Atmospheric Optics

The unique optical properties of the atmosphere cause a wide range of spectacular optical phenomena. The blue colour of the sky is a direct result of Rayleigh scattering which redirects higher frequency (blue) sunlight back into the field of view of the observer. Because blue light is scattered more easily than red light, the sun takes on a reddish hue when it is observed through a thick atmosphere, as during a sunrise or sunset. Additional particulate matter in the sky can scatter different colours at different angles creating colourful glowing skies at dusk and dawn. Scattering off of ice crystals and other particles in the atmosphere are responsible for halos, afterglows, coronas,rays of sunlight, and sun dogs. The variation in these kinds of phenomena is due to different particle sizes and geometries.

Figure: A colourful sky is often due to scattering of light off particulates and pollution, as in this photograph of a sunset during the October 2007 California wildfires.

Mirages are optical phenomena in which light rays are bent due to thermal variations in the refraction index of air, producing displaced or heavily distorted images of distant objects.

Other dramatic optical phenomena associated with this include the Novaya Zemlya effect where the sun appears to rise earlier than predicted with a distorted shape. A spectacular form of refraction occurs with a temperature inversion called the Fata Morgana where objects on the horizon or even beyond the horizon, such as islands, cliffs, ships or icebergs, appear elongated and elevated, like "fairy tale castles".

Rainbows are the result of a combination of internal reflection and dispersive refraction of light in raindrops. A single reflection off the backs of an array of raindrops produces a rainbow with an angular size on the sky that ranges from 40° to 42° with red on the outside.

Double rainbows are produced by two internal reflections with angular size of 50.5° to 54° with violet on the outside. Because rainbows are seen with the sun 180° away from the centre of the rainbow, rainbows are more prominent the closer the sun is to the horizon.

Some Common Terms—Light

Light or visible light is electromagnetic radiation that is visible to the human eye, and is responsible for the sense of sight. Visible light has wavelength in a range from about 380 nanometres to about 740 nm, with a frequency range of about 405 THz to 790 THz.

In physics, the term *light* sometimes refers to electromagnetic radiation of any wavelength, whether visible or not. Primary properties of light are intensity, propagation direction, frequency or wavelength spectrum, and polarisation, while its speed in a vacuum, 299,792,458 metres per second (about 300,000 kilometres per second), is one of the fundamental constants of nature.

Light, which is emitted and absorbed in tiny "packets" called photons, exhibits properties of bothwaves and particles. This property is referred to as the wave–particle duality. The study of light, known as optics, is an important research area in modern physics.

Speed of Light

The speed of light in a vacuum is defined to be exactly 299,792,458 m/s (approximately 186,282 miles per second). The fixed value of the speed of light in SI units results from the fact that the metre is now defined in terms of the speed of light. Different physicists have attempted to measure the speed of light throughout history.

Galileo attempted to measure the speed of light in the seventeenth century. An early experiment to measure the speed of light was conducted by Ole Rψmer, a Danish physicist, in 1676. Using a telescope, Rψmer observed the motions of Jupiter and one of its moons, Io. Noting discrepancies in the apparent period of Io's orbit, he calculated that light takes about 22 minutes to traverse the diametre of Earth's orbit. Unfortunately, its size was not known at that time.

If Rψmer had known the diametre of the Earth's orbit, he would have calculated a speed of 227,000,000 m/s. Another, more accurate, measurement of the speed of light was performed in Europe by Hippolyte Fizeau in 1849. Fizeau directed a beam of light at a mirror several kilometres away. A rotating cog wheel was placed in the path of the light beam as it travelled from the source, to the mirror and then returned to its origin. Fizeau found that at a certain rate of rotation, the beam would pass through one gap in the wheel on the way out and the next gap on the way back. Knowing the distance to the mirror, the number of teeth on the wheel, and the rate of rotation, Fizeau was able to calculate the speed of light as 313,000,000 m/s.

Léon Foucault used an experiment which used rotating mirrors to obtain a value of 298,000,000 m/s in 1862. Albert A. Michelson conducted experiments on the speed of light from 1877 until his death in 1931. He refined Foucault's methods in 1926 using improved rotating mirrors to measure the time it took light to make a round trip from Mt. Wilson to Mt. San Antonio in California.

The precise measurements yielded a speed of 299,796,000 m/s. Two independent teams of physicists were able to bring light to a complete standstill by passing it through a Bose-Einstein Condensate of the element rubidium, one team led by Dr.

Lene Vestergaard Hau of Harvard University and the Rowland Institute for Science in Cambridge, Mass., and the other by Dr. Ronald L. Walsworth and Dr. Mikhail D. Lukin of the Harvard-Smithsonian Centre for Astrophysics, also in Cambridge.

Optics

The study of light and the interaction of light and matter is termed optics. The observation and study of optical phenomena such as rainbows and the aurora borealis offer many clues as to the nature of light as well as much enjoyment.

Refraction

Figure: An example of refraction of light. The straw appears bent, because of refraction of light as it enters liquid from air.

Refraction is the bending of light rays when passing through a surface between one transparent material and another. It is described by Snell's Law:

$$n_1 \sin\theta_1 = n_2 \sin\theta_2 \, .$$

where θ_1 is the angle between the ray and the surface normal in the first medium, θ_2 is the angle between the ray and the surface normal in the second medium, and n_1 and n_2 are the indices of refraction, $n = 1$ in a vacuum and $n > 1$ in a transparent substance.

When a beam of light crosses the boundary between a vacuum and another medium, or between two different media, the wavelength of the light changes, but the frequency remains constant. If the beam of light is not orthogonal (or rather normal) to the boundary, the change in wavelength results in a change in the direction of the beam. This change of direction is known as refraction.

The refractive quality of lenses is frequently used to manipulate light in order to change the apparent size of images. Magnifying glasses, spectacles, contact lenses, microscopes and refracting telescopesare all examples of this manipulation.

Light Sources

Figure: A cloud illuminated by sunlight

There are many sources of light. The most common light sources are thermal: a body at a given temperature emits a characteristic spectrum of black-body radiation. Examples include sunlight (the radiation emitted by the chromosphere of the Sun at around 6,000 Kelvin peaks in the visible region of the electromagnetic spectrum when plotted in wavelength units and roughly 40% of sunlight is visible), incandescent light bulbs (which emit only around 10% of their energy as visible light and the remainder as infrared), and glowing solid particles in flames. The peak of the blackbody spectrum is in the infrared for relatively cool objects like human beings. As the temperature increases, the peak shifts to shorter wavelengths, producing first a red glow, then a white one, and finally a blue colour as the peak moves out of the visible part of the spectrum and into the ultraviolet. These colours can be seen when metal is heated to "red hot" or "white hot". Blue thermal emission is not often seen. The commonly seen blue colour in a gas flame or a welder's torch is in fact due to molecular emission, notably by CH radicals (emitting a wavelength band around 425 nm).

Atoms emit and absorb light at characteristic energies. This produces "emission lines" in the spectrum of each atom. Emission can be spontaneous, as in light-emitting diodes, gas discharge lamps (such as neon lamps and neon signs, mercury-vapour lamps, etc.), and flames (light from the hot gas itself—so, for example, sodium in a gas flame emits characteristic yellow light). Emission can also be stimulated, as in a laser or a microwave maser. Deceleration of a free charged particle, such as an electron, can produce visible radiation: cyclotron radiation, synchrotron radiation, and brems strahlung radiation are all examples of this. Particles moving through a medium faster than the speed of light in that medium can produce visible Cherenkov radiation. Certain chemicals produce visible radiation by chemoluminescence. In living things, this process is called bioluminescence. For example, fireflies produce light by this means, and boats moving through water can disturb plankton which produce a glowing wake. Certain substances produce light when they are illuminated by more energetic radiation, a process known as fluorescence. Some substances emit light slowly after excitation by more energetic radiation. This is known as phosphorescence.

Phosphorescent materials can also be excited by bombarding them with subatomic particles. Cathodoluminescence is one example. This mechanism is used in cathode ray tube television sets and computer monitors.

Certain other mechanisms can produce light:

- Scintillation
- Electroluminescence
- Sonoluminescence
- Triboluminescence
- Cherenkov radiation
- Bioluminescence.

When the concept of light is intended to include very-high-energy photons (gamma rays), additional generation mechanisms include:

- Radioactive decay
- Particle–antiparticle annihilation.

Figure: A city illuminated by artificial lighting

Light Pressure

Light exerts physical pressure on objects in its path, a phenomenon which can be deduced by Maxwell's equations, but can be more easily explained by the particle nature of light: photons strike and transfer their momentum. Light pressure is equal to the power of the light beam divided by *c*, the speed of light.

Due to the magnitude of *c*, the effect of light pressure is negligible for everyday objects. For example, a one-milliwatt laser pointer exerts a force of about 3.3 piconewtons on the object being illuminated; thus, one could lift a U. S. penny with laser pointers, but doing so would require about 30 billion 1-mW laser pointers. However, in nanometre-scale applications such as NEMS, the effect of light pressure is more pronounced, and exploiting light pressure to drive NEMS mechanisms and to flip nanometre-scale physical switches in integrated circuits is an active area of research.

At larger scales, light pressure can cause asteroids to spin faster, acting on their irregular shapes as on the vanes of a windmill. The possibility to make solar sails that would accelerate spaceships in space is also under investigation. Although the motion of the Crookes radiometre was originally attributed to light pressure, this interpretation is incorrect; the characteristic Crookes rotation is the result of a partial vacuum. This should not be confused with the Nichols radiometre, in which the motion *is* directly caused by light pressure.

Closed-eye Hallucination

Closed-eye hallucinations and closed-eye visualisations (CEV) are a distinct class of hallucination. These types of hallucinations generally only occur when one's eyes are closed or when one is in a darkened room. They are a form of phosphene. Certain users report closed-eye hallucinations under the influence of psychedelics. These are reportedly of a different nature than the "open-eye" hallucinations of the same compounds.

Levels of CEV Perception

There are five known levels of CEV perception which can be achieved either through chemical stimuli or through meditative relaxation techniques.

Level 1: Visual Noise

The most basic form of CEV perception that can be immediately experienced in normal waking consciousness involves a seemingly random noise of pointillistic light/dark regions with no apparent shape or order. This can be seen when the eyes are closed and looking at the back of the eyelids. In a bright room, a dark red can be seen, owing to a small amount of light penetrating the eyelids and taking on the colour of the blood within them. In a dark room, blackness can be seen or the object can be more colourful. But in either case it is not a flat unchanging redness/blackness. Instead, if actively observed for a few minutes, one becomes aware of an apparent disorganised motion, a random field of lightness/darkness that overlays the redness/blackness of your closed eyelids.

For a person who tries to actively observe this closed-eye perception on a regular basis, there comes a point where if they look at a flat-shaded object with their eyes wide open, and try to actively look for this visual noise, they will become aware of it and see the random pointillistic disorganised motion as if it were a translucent overlay on top of what is actually being seen by their open eyes. When seen overlaid onto the physical world, this CEV noise does not obscure physical vision at all, and in fact is hard to notice if the visual field is highly patterned, complex, or in motion. When active observation is stopped, it is not obvious or noticeable, and seemingly disappears from normal physical perception. Individuals suffering from visual snow see similar noise but experience difficulty blocking it from conscious perception. The noise probably originates from thermal noise exciting the photoreceptor cells in the retina, compare Eigengrau.

Level 2: Light/dark Flashes

Some mental control can be exerted over these closed-eye visualisations, but it usually requires a bit of relaxation and concentration to achieve.

When properly relaxed it is possible to cause regions of intense black, bright white or even colours such as yellow, green, or pink to appear in the noise. These regions can span the entire visual field, but seem to be fleeting in nature.

Level 3: Patterns, Motion, and Colour

This level is relatively easily accessible to people who use LSD, and appears to be what most people refer to as colourful visuals. However, it is also accessible to people involved in deep concentration for long periods of time. When lying down at night and closing the eyes, right before sleep the complex motion of these patterns can become directly visible without any great effort thanks to hypnagogic hallucination. The patterns themselves might resemble fractals.

Level 4: Objects and Things

This is a fairly deep state. At this level, thoughts visually manifest as objects or environments. When this level is reached, the CEV noise seems to calm down and fade away, leaving behind an intense flat ordered blackness. The visual field becomes a sort of active space. A side component of this is the ability to feel motion when the eyes are closed.

Opening the eyes returns one to the normal physical world, but still with the CEV object field overlaid onto it and present. In this state it is possible to see things that appear to be physical objects in the open-eye physical world, but that aren't really there.

> *"If we remember that the essential difference between what we call the real world and the world of imagination and hallucination, is not the elements of which we build them up but the sequence in which these elements appear... then it follows that the sequences directed from without represent a limitation of the otherwise unlimited combinations of the selective forms released at random from within" —Jurij Moskvitin, Essay on the origin of thought.*

Level 5: Overriding Physical Perception

This is the point where it appears to the outside world that a person is either unconscious or insane. The internal CEV perceptions and think-it/feel-it perceptions become stronger than physical perceptions, and completely override and replace open-eye physical perceptions. This can be a potentially dangerous state if a person is still mobile while in a different imagined world, but by this time most people are motionless and not likely to do something hazardous to themselves or others. This is the point where most hallucinogenic references say it is a good idea to have a "sitter" present to watch over the person using the chemicals, and keep them from accidentally harming themselves or others while deep into their own world.

What is not a CEV

Image burn-in (afterimage)

Image burn-in occurs when very bright objects lie in one's field of vision, and should not be confused with closed-eye hallucinations. Visual burn-in from bright lights is visible for a few minutes after closing the eyes, or by blinking repeatedly, but the burn-in effect slowly fades away as the retina recovers, whereas the waking-consciousness CEV noise will not disappear if observed continuously over a period of time.

Corneal Liquid

CEV does not involve the liquid and air bubbles on the surface of the cornea, which can also be seen by extremely nearsighted people when looking at bright point-light sources with glasses/contacts removed. Also called "Floaters" - often appear as cells floating across the eye. Half-closing and reopening the eyelids creates a very definite wiper-ridge in the corneal liquid that is readily visible. Fully closing and reopening the eyelids also stirs up the corneal liquid which settles down after a brief moment. The motion of waking-consciousness CEV noise is not so directly and physically controllable and repeatable. This is not necessarily only associated with extreme nearsightedness.

Blue-sky Sprites

CEV does not seem to be related to the "sprites" (blue field entoptic phenomenon) that can be seen as dots darting around when staring up into a bright blue sky on a sunny day (not looking at the sun). These dots superimposed over a flat blue background are white blood cells moving through the blood vessels of the retina. The motion of waking-consciousness CEV noise is uniformly random compared to the waking-consciousness blue-sky sprite motion.

Physical Retinal Stimulation

CEV is unrelated to the visual noise seen when the retina is physically stimulated. The retina can be made to produce light patterns of visual noise simply by one rubbing their eyes somewhat forcefully in a manner that increases intraocular pressure. Additionally retinal noise can be produced by touching near the rear of the eyeball (for example, if one closes their eyes, looks all the way left, and lightly touches the rightmost part of the eye socket, this produces visual noise in the shape of a circle that appears at the left side of the visual field - a practice that is neither painful nor dangerous). None of these are closed-eye hallucinations.

Monocular

A monocular is a modified refracting telescope used to magnify the images of distant objects by passing light through a series of lenses and sometimes prisms; the use of prisms results in a lightweight telescope. Volume and weight are less than half those of binoculars of similar optical properties, making it easy to carry. Monoculars produce 2-dimensional images, while binoculars add perception of depth (3 dimensions).

A monocular with a straight optical path is relatively long; prisms can be used to fold the optical path to make an instrument which is much shorter. Monoculars, sometimes called *telescopes* when used in this capacity, are used wherever a magnified 2-dimensional image of a distant object is required.

Visually impaired people may use monoculars to see objects at distances at which people with normal vision do not have difficulty, e.g., to read text on a chalkboard or projection screen. Applications for viewing more distant objects include natural history, hunting, marine and military applications.

When high magnification, a bright image, and good resolution of distant images are required, a relatively large instrument is preferred, often mounted on a tripod. Smaller pocket-sized "pocket scopes" can be used for less stringent applications.

Phosphene

A phosphene is a phenomenon characterised by the experience of seeing light without light actually entering the eye. The word *phosphene* comes from the Greek words *phos* (light) and *phainein* (to show). Phosphenes are flashes of light, often associated with optic neuritis, induced by movement or sound. Phosphenes can be directly induced by mechanical, electrical, or magnetic stimulation of the retina or visual cortex as well as by random firing of cells in the visual system. Phosphenes have also been reported by meditators (commonly called *nimitta*); people who go for long periods without visual stimulation (also known as the prisoner's cinema); or those who are using psychedelic drugs.

Causes

Mechanical Stimulation

The most common phosphenes are pressure phosphenes, caused by rubbing the closed eyes. They have been known since antiquity, and described by the Greeks. The pressure mechanically stimulates the cells of the retina. Experiences include a darkening of the visual field that moves against the rubbing, a diffuse coloured patch that also moves against the rubbing, a scintillating and ever-changing and deforming light grid with occasional dark spots (like a crumpling fly-spotted flyscreen), and a sparse field of intense blue points of light. Pressure phosphenes can persist briefly after the rubbing stops and the eyes are opened, allowing the phosphenes to be seen on the visual scene. Hermann von Helmholtz and others have published drawings of their pressure phosphenes. One example of a pressure phosphene is demonstrated by gently pressing the side of your eye and observing a coloured ring of light on the opposite side, as detailed by Isaac Newton.

Another common phosphene is "seeing stars," from a sneeze, a heavy and deep cough, blowing of the nose, a blow on the head or low blood pressure (such as on standing up too quickly or prior to fainting). It is possible these involve some mechanical stimulation of the retina, but they may also involve mechanical and metabolic stimulation of neurons of the visual cortex or of other parts of the visual system.

Less commonly, phosphenes can also be caused by some diseases of the retina and nerves, for example multiple sclerosis. The British National Formulary lists phosphenes as an occasional side effect of at least one anti-anginal medication (available only by medical prescription in the United Kingdom). The name "phosphene" was coined by JBH Savigny, better known as the ship's surgeon of the wrecked *Medusa*. It was first employed by Serre d'Uzes to test retinal function prior to cataract surgery.

Electrical Stimulation

Phosphenes have also been created by electrical stimulation of the brain as early as 1929 by neurologist Otfrid Foerster. Brindley and Lewin (1968) inserted a matrix of stimulating electrodesdirectly into the visual cortex of a 52-year-old blind man, using small pulses of electricity to create phosphenes. These phosphenes were points, spots, and bars of colourless or coloured light. Brindley and Rushton (1974) used the phosphenes to create a visual prosthesis, in this case by using the phosphenes to depict Braille spots.

In recent years, researchers have successfully developed experimental brain-computer interfaces or neuroprosthesis that stimulate phosphenes to restore vision to people blinded through accidents. Notable successes include the human experiments by William H. Dobelle and Mark Humayun and animal research by Dick Normann.

Experiments with humans have shown that when the visual cortex is stimulated above the calcarine fissure phosphenes are produced in the lower part of the visual field, and vice versa.

Other Causes

Phosphenes have also been created by intense, changing magnetic fields, such as with transcranial magnetic stimulation. These fields can be positioned on different parts of the head to stimulate cells in different parts of the visual system. They also can be induced by alternating currents that entrainneural oscillation as with trancranial alternating-current stimulation. In this case they appear in the peripheral visual field.

Astronauts exposed to radiation in space report seeing phosphenes.

Explanation

Most vision researchers believe that phosphenes result from the normal activity of the visual system after stimulation of one of its parts from some stimulus other than light. For example, Grüsser et al. showed that pressure on the eye results in activation of retinal ganglion cells in a similar way to activation by light.

An ancient, discredited theory is that light is generated in the eye. A version of this theory has been revived, except, according to its author, that "phosphene lights are [supposed to be] due to the intrinsic perception of induced or spontaneous increased biophoton emission of cells in various parts of the visual system (from retina to cortex)"

Anthropological Research

In 1988, David Lewis-Williams and T. A. Dowson published an article about phosphenes and other entoptic phenomena. They argued, among other things, that non-figurative art of the Upper Paleolithic depicts actual visions of phosphenes and neurological "form constants," probably enhanced by hallucinogenic drugs. Phosphenes were used as a learning technique by French scientist and scholar, Francis Lefebure this is called "phosphenism".

Diphosphene

Diphosphene is a compound having the formula $(PH)_2$. It exists as two geometric isomers, *E* and *Z*. Diphosphene is also the parent member of the entire class of diphosphene compounds with the formula $(PR)_2$, where R is an organyl group.

Visual Snow

Visual snow is a transitory or persisting visual symptom where people see snow or television-likestatic in parts or the whole of their visual fields, especially against dark backgrounds. It is much like camera noise in low light conditions.

The severity or density of the "snow" differs from one person to the next; in some circumstances, it can inhibit a person's daily life, making it difficult to read, see in detail and focus correctly. The "snow" is more generalised than the "blue-sky sprites" seen in the blue field entoptic phenomenon.

No etiology for visual snow has been identified, and anecdotal reports point to a multitude of associated conditions, possibly rendering it a non-specific symptom. Insofar as sufferers of visual snow have undergone ophthalmic, neurological and psychiatric examinations, no systematic problems besides the visual snow have been identified. Pending recognition of the condition, little medical research is taking place to possibly identify more subtle deviations.

Causes

Visual snow can occur in a variety of ophthalmic disorders that can be diagnosed by the presence of additional clinical signs and symptoms. Persisting visual snow can feature as a leading symptom of a migraine complication called persistent aura without infarction, commonly referred to as persistent migraine aura (PMA). It is important to keep in mind that there exist many clinical sub-forms of migraine where headache may be absent and where the migraine aura may not take the typical form of the zigzagged fortification spectrum, but manifests with a large variety of focal neurological symptoms. A condition that sometimes produces visual snow is optic neuritis (inflammation of the optic nerve), caused by multiple sclerosis (MS). Moreover, a variety of illnesses (e.g. Lyme disease, auto-immune disease) or noxious events (e.g. prolonged use of a VDU, dehydration, over-acidification) have been blamed by sufferers in self-help internet forums as causes of persisting visual snow, but none of these claims have been confirmed by scientific study. Some patients fail to find any apparent causative illness or event in their lives, instead saying the snow came out of nowhere or has been with them for their whole life.

Hallucinogen persisting perception disorder (HPPD) is another condition which has resulted in the onset of the visual snow following the use of hallucinogenic psychedelic drugs. In HPPD, the symptom of the visual disturbances has been described as aeropsia (literally "seeing the air"). HPPD very rarely occurs after just a single dose of a hallucinogenic drug and with a considerable latency between last drug intake and onset of persistent perception disorder, so taking a thorough life-time drug history is mandatory in the diagnostic work-up of visual snow. In many cases, the neurological action for HPPD is not known, and the majority of evidence surrounding it is anecdotal and difficult to isolate.

Related Symptoms

In addition to visual snow, many sufferers have other types of visual disturbances such as starbursts, increased afterimages, floaters, trails, and many others. Non-visual symptoms such as tinnitus, depersonalisation-derealisation, fatigue, speech difficulties and cognitive dysfunction (brain fog) are frequently encountered. Secondary psychiatricsequelae such as anxiety, panic attacks or depression may develop and necessitate appropriate treatment.

Treatments

There currently is no established treatment for visual snow. In HPPD, clonazepam has been recommended as medication of first choice in patients seeking medical help. Furthermore, drug abstinence is sometimes said to be of major therapeutic importance in HPPD. In persistent aura without infarction, the evidence so far suggests that acetazolamide may be the premier drug for patients with the repetitive form of aura status and that valproate, lamotrigine, or topiramateshould be first choices for patients with the continuous form. When these oral drugs are ineffective, an intravenous injection or injections of furosemide should be tried. However, with very little scientific research on the condition taking place, for the time being the effectiveness of such treatments remains based solely on anecdotal evidence. Beyond pharmacological approaches, appropriate counselling and cognitive behavioural interventions that focus on coping with the condition may be of huge importance.

Scintillating Scotoma

Scintillating scotoma is the most common visual aurapreceding migraine and was first described by 19th century physician Hubert Airy (1838–1903). It is often confused with ocular migraine which originates in the eyeball or socket.

Presentation

Figure: Artist's depiction of scintillating scotoma

Figure: Example of a scintillating scotoma, as may be caused by cortical spreading depression.

It may occur as an isolated symptom without headache in acephalgic migraine. Although many variations occur, scintillating scotoma usually begins as a spot of flickering light near or in the centre of the visual fields, which prevents vision within the scotoma. The scotoma area flickers, but is not dark. The scotoma then expands into one or more shimmering arcs of white or coloured flashing lights. An arc of light may gradually enlarge, become more obvious, and may take the form of a definite zigzag pattern, sometimes called a fortification spectrum, because of its resemblance to the fortifications of a castle or fort seen from above. Patterns within the arc have also been known to share striking similarity with Widmanstätten patterns. The scotoma may be bilateral or unilateral and vision beyond the borders of the expanding scotoma(s) remains normal.

The visual anomaly results from abnormal functioning of portions of the occipital cortex, at the back of the brain, not in the eyes nor any component thereof, such as the retinas. This is a different disease from retinal migraine, which is monocular (only one eye).

It may be difficult to read and dangerous to drive a vehicle while the scotoma is present. Normal central vision may return several minutes before the scotoma disappears from peripheral vision.

Sufferers can keep a diary of dates on which the episodes occur to show to their physician, plus a small sketch of the anomaly, which may vary between episodes.

Causes

There is no generally accepted cause of scintillating scotomas, but monosodium glutamate (MSG) is frequently reported as a dietary trigger. According to the National Library of Medicine's Medical Encyclopedia, migraine attacks may be triggered by:

- Allergic reactions
- Bright lights
- Loud noises
- Certain odors
- Perfumes
- Physical or emotional stress
- Changes in sleep patterns
- Smoking or exposure to smoke
- Skipping meals
- Alcohol
- Caffeine
- Menstrual cycle fluctuations

- Birth control pills
- Tension headaches
- Foods containing tyramine (red wine, aged cheese, smoked fish, chicken livers, figs, and some beans)
- Monosodium glutamate (MSG)
- Nitrates (like bacon, hot dogs, and salami)
- Other foods such as chocolate, nuts, peanut butter, avocado, banana, citrus, onions, dairy products, and fermented or pickled foods.

Prognosis

Symptoms typically appear gradually over 5 to 20 minutes and generally last fewer than 60 minutes, leading to the headache in classic migraine with aura, or resolving without consequence in acephalgic migraine. Many migraine sufferers evolve from scintillating scotoma as a prodrome to migraine, to scintillating scotoma without migraine. The scotoma typically spontaneously resolves within the stated time frame, leaving few or no subsequent symptoms, though some report fatigue, nausea and dizziness as sequelae.

Dark Retreat

Dark retreat refers to advanced practices in the Dzogchen lineages of the Nyingmapa, Bφnpo and other schools of Tibetan Buddhism. The time period dedicated to dark retreat varies from a few hours to decades. Dark Retreat in the Himalayan tradition is a restricted practice only to be engaged by the senior spiritual practitioner under appropriate spiritual guidance. This practice is considered conducive for navigating the bardo at the time of death and for realising the rainbow body. The traditional dark retreat requires stability in the natural state and is only suitable for advanced practitioners.

Ayu Khandro and Dilgo Khyentse are examples of modern, if not contemporary, practitioners of significant periods of Dark Retreat sadhana. Ayu Khandro performed the *Yang-Ti* (Tibetan), an advanced practice of the Dzogchen Upadesha, a version of the Dark Retreat.

Dark Retreat in other Cultures

Ancient Egyptians and Mayans practiced a form of the dark retreat, traditionally lasting ten days. Holy men would enter into the centre of their respective pyramids, completely removed from light and sound, and have visions of the workings of the universe.

Isolation Tank

An isolation tank is a lightless, soundproof tank inside which subjects float in salt water at skin temperature. They were first used by John C. Lilly in 1954 to test the effects of sensory deprivation. Such tanks are now also used for meditation and relaxation and in alternative medicine. The isolation tank was originally called the sensory deprivation tank. Other names for the isolation tank include flotation tank, John C. Lilly tank, REST tank, sensory attenuation tank, and think tank.

History

John C. Lilly, a medical practitioner and neuro-psychiatrist, developed the flotation tank in 1954. During his training in psychoanalysis at the US National Institute of Mental Health (NIMH), Lilly commenced experiments with sensory deprivation. In neurophysiology, there had been an open question as to what keeps the brain going and the origin of its energy sources. One hypothesis was that the energy sources are biological and internal and do not depend upon the outside environment. It was argued that if all stimuli are cut off to the brain then the brain would go to sleep. Lilly decided to test this hypothesis and, with this in mind, created an environment which totally isolated an individual from external stimulation. From here, he studied the origin of consciousness and its relation to the brain.

Peter Suedfeld and Roderick Borrie of the University of British Columbia began experimenting on the therapeutic benefits of flotation tank usage in the late 1970s. They named their technique "Restricted Environmental Stimulation Therapy" (REST).

Tank Design and Usage

In the original tanks, people were required to wear complicated head-masks to breathe underwater, but the mask detracted from the isolation experience. The tight fit of the mask seal around the face and the retention strips wrapping around the back of the head were uncomfortable in long sessions. The constant hissing of the air valves and bubbling of exhaust air out of the mask prevented the possibility of silence. The faceplate of the mask was typically solid black for visual isolation, but then the tank user needed help entering and leaving the isolation tank since the mask blinded them.

In newer tanks, epsom salt is added to the water in the tank to increase the density of the water so that the subject floats with the face above the water. However, since the ears are submerged when the subject is in a relaxed position, hearing is greatly reduced, particularly when ear-plugs are also used. When the arms float to the side, skin sensation is greatly reduced because the air and water are the same temperature as the skin, and the feeling of a body boundary fades. The sense of smell is also greatly reduced, especially if the water has not been treated with chlorine.

The growing number of commercial float tanks has brought increased regulation of disinfection. For example, different states in the US have applied different rules. In Europe, the DIN 19643 bathing water standard requires automated chlorination from REDOX (ORP) measurement. Chlorine, bromine and peroxide disinfection have all been used successfully. There is doubt about ozone safety because of the closed air space.

Generally, users of isolation tanks enter the pool nude. Although a swimsuit may be worn, the elastic material of a tight-fitting suit can create uncomfortable compressed stress points on the skin during the session. Due to the high epsom salt content, the water is minimally changed, and all users are expected to shower, soap up, and rinse clean prior to entering the tank to extend the water life as long as possible. Bathing is again needed after a session to remove excess epsom salt from the skin.

Most isolation tanks use a circulating surface skimmer, cartridge filtration, and disinfection means, including ultraviolet sterilization and chemicals to keep the water free of microbes and sediment, though this is turned off during a session to keep the isolation space as quiet as possible. A ring heating system can be used around the outer walls of the tank to warm the water so that it rises up the outside edges of the pool, travels towards the centre, and then sinks under the tank user.

This very slow water convection flow helps to keep the user centred in the middle of the pool, without them floating to the side and bumping into the walls of the small tank during long float sessions. However, when the floater is still, the small waves caused by breathing also tend to centre them in the solution regardless of a convection effect.

Isolation tank construction and plumbing is typically all-plastic. High quality flotation tanks may use medical stainless steel which is impervious to the strong salt solution and more importantly the disinfectants, lower quality tanks use lesser metals which will corrode. Unsealed stone and concrete surfaces can also be damaged by dissolved Epsom salt splashed or dripped outside the tank, due to the dried salt recrystallizing and opening up cracks and fissures.

It is useful to have a shower stall in the same room as the tank. Before a session, the tank user can transfer directly from the shower to the tank without drying off. Following the session, they can immediately rinse off which minimises the dripping of corrosive salt through the building.

Having plumbing facilities immediately next to the tank is also helpful when the water must eventually be changed to prevent microbe growth. The plumbing, including the drain pipes, should be constructed of plastic to prevent deterioration from the salt. The salt concentration may need to be diluted when discarded, to prevent damage to small private wastewater plumbing systems.

Types of Sessions

A therapeutic session in a flotation tank typically lasts an hour. For the first forty minutes, it is reportedly possible to experience itching in various parts of the body (a phenomenon also reported to be common during the early stages of meditation). The last twenty minutes often end with a transition from beta or alpha brainwaves to theta, which typically occurs briefly before sleep and again at waking. In a float tank, the theta state can last for several minutes without the subject losing consciousness. Many use the extended theta state as a tool for enhanced creativity and problem solving or for superlearning. The more often the tank is used the longer the theta period becomes. Spas sometimes provide commercial float tanks for use in relaxation. Flotation therapy has been academically studied in the USA and in Sweden with published results showing reduction of both pain and stress. The relaxed state also involves lowered blood pressure and maximal blood flow.

Floating can be passive or active, depending on the purpose. For relaxation, one simply floats and 'clears the mind.' Active floating has many different techniques. One may perform meditation, mantras, self-hypnosis, utilise educational programs, etc. The idea of active floating is that, when the body is relaxed, the mind becomes highly suggestible and any action taken during these states will enter the information into the sub-conscious. Flotation therapy may be used to complement other body work and healing methods.

More extreme uses of the tank involve the subject taking varying doses of psychedelics; such as LSD, psilocybin mushrooms, dimethyltryptamine, ketamine, or cannabis and spending prolonged periods in the tank (up to tens of hours) at a time, an approach pioneered by John C. Lilly himself—though he claims to have tried LSD in the tank only prior to 1964, while the psychedelic was still legal.. This is regarded by commercial float centres as extremely dangerous behaviour and is completely prohibited.

Flotation Therapy

Flotation therapy is therapy that is undertaken by floating in a warm salt water in a float tank.

Flotation therapy developed from the research work of John Lilly although he was not primarily interested in therapy, rather in the effect of sensory deprivation on the human brain and mind.

People using early float tanks discovered that they enjoyed the experience and that the relaxed state was also a healing state for many conditions including stress, anxiety, pain, swelling, insomnia and jet lag.

As a result float tanks were produced for commercial uses and commercial float centres offering flotation therapy opened in several countries during the period from 1980 to the present day when there are hundreds of flotation centres in dozens of countries. In almost all cases these float centres offer wellness treatments and in particular the release of stress.

Research into flotation therapy (as opposed to just the effect of isolation) began in the USA at Ohio State University where floating was shown to improve creativity in Jazz musicians, accuracy in rifle shooting, focus before academic examinations and stress relief, among others.

Research in Sweden has demonstrated the therapeutic effect on stress and pain.

The technique takes advantage of an atavistic ability that seems to be common to all humans to relax when floating at a comfortable temperature. The temperature is that which allows natural heat generation to escape without the need for muscle action to raise body temperature in homeostasis. The floating posture, usually the supine position (although the prone position with chin supported on elbows is recommended for pregnant women), allows all the postural muscles to relax. The water pressure on the immersed skin is lower than the blood pressure and thus blood flow continues in skin capillaries. This is in contrast to normal bed rest where local contact pressure inhibits blood flow requiring regular adjustment of posture. When people cannot adjust their posture in bed, e.g. in some illnesses, bed sores can result. When floating there is no tendency to adjust posture and a person can float immobile for many hours.

The natural tendency of the body in the floating posture at the correct temperature is to dilate the blood vessels, reducing the blood pressure and maximising blood flow. The brain activity normally associated with postural muscles is reduced to a minimum. In this state, which we can call the floating state, natural endorphins are released reducing pain. Lactic acid removal is accelerated. Flow in the lymphatic system is increased.

Effects

The Effect on Stress

Perceived stress can be correlated with increased levels of cortisol and in flotation therapy there is a natural tendency for cortisol to be reduced. For this reason, floatation therapy is one of the few noninvasive techniques available to manage stress when it is a factor in reducing a person's ability to cope with normal life. Flotation therapy is a fast technique in this respect. The Swedish research was based on 40 minute float sessions. This compares well with other management techniques such as long vacations.

There are many similarities with the age old long hot bath, the differences being that in floatation therapy the temperature is maintained at the correct level and the bath is large enough to float without touching the sides of the bath.

The Effect of the Salt

Most float tanks use epsom salt, Magnesium sulfate, in high concentration so that the relative density of the solution is about 1.25. (Lilly recommended 1.3 but this requires operating very close to saturation with the risk of recrystalisation). The density assists floating particularly making the head buoyant so that the nose and mouth are well out of the water for breathing.

It has recently been discovered that there is a secondary effect which is important to flotation therapy. Magnesium is absorbed through the skin due to natural molecular diffusion. This tends to correct a common deficiency. Magnesium is absorbed from the diet but in many areas of the world, over-cropping without adequate replacement of magnesium makes the normal diet low in magnesium. The body naturally optimises the levels of magnesium, so there is no overload effect from floating in the salts for extended periods. It is noted that there is no "flotation therapist", although there is a need to instruct the floater and need to maintain the equipment in a safe condition. However, flotation therapy is compatible with other therapies as a preparation or conjuct activity. Examples include massage, talk therapy, and hypnosis.

Examples of Floatation Therapy

Fertility: Many women, after medical examination for physical causes, are advised to take significant lifestyle changes such as moving house or stopping work because stress is recognised as a factor in infertility. Flotation therapy is a useful stress reduction technique in many cases.

Insomnia: Loss of sleep or disturbed sleep patterns can exacerbate stress and flotation therapy is an ideal technique to reduce stress in anticipation of bed rest. Many floaters fall asleep in the float tank. This is considered safe in that there is no tendency to roll over when floating. Long periods of sleep are not usually possible in a commercial float centre but the therapy has long lasting effects so the sufferer can go to bed after a float session and sleep naturally without drugs.

Definition

Latest Research

New research undertaken at the Human Performance Laboratory at Karlstad University Sven-Åke Bood concludes that regular flotation tank sessions can provide significant relief for chronic stress-related ailments. Studies involving 140 people with long-term conditions such as anxiety, stress, depression and fibromyalgia found that more than three quarters experienced noticeable improvements.

Dr. Bood commented: "Through relaxing in floating tanks, people with long-term fibromyalgia, for instance, or depression and anxiety felt substantially better after only 12 treatments". Research targeted the effectiveness of floatation treatment with regard to stress related pain and anxiety over the period of seven weeks. 22 percent of the participants became entirely free of pain and 56 percent experienced clear improvement.

Broken down to various symptoms, the results were as follows: 23 percent slept better, 31 percent experienced reduced stress, 27 percent felt less agony and 24 percent became less depressed or got rid of their depression altogether. The research also confirms the findings of an earlier thesis that flotation, after only twelve sessions, substantially improves sleep patterns leaving users more optimistic and with reduced nervousness, tension and pain. Relaxing in a weightless state in the silent warmth of a flotation tank activates the body's own system for recuperation and healing, said Sven-Åke Bood. What researchers find particularly gratifying is that the positive effects were still in evidence 4 months after the floating treatment ended.

Notable Users

The physicist Richard Feynman's experiences in a sensory deprivation tank were documented in the popular book Surely You're Joking, Mr. Feynman!. Feynman was invited to try the isolation tank at John Lilly's home after Lilly attended one of Feynman's popular lectures on quantum mechanics. Comedian Joe Rogan owns an isolation tank and can be heard talking about his experiences with it in a YouTube video, in which he was giving away his previous model as he had recently purchased a larger, more intricate system. Joe exclaimed his experiences with isolation tanks were beneficially catalysed when he ingested cannabis before entering the tank. Another comedian, George Carlin used an isolation tank to "meditate or just drift off".

Actor Nicolas Cage spent some time in an isolation tank to capture the claustrophobia of his character in the 2006 film World Trade Centre. He played PAPD Sgt. John McLoughlin, who was trapped in the rubble of the WTC Towers during the attacks.

Francis Lefebure

Dr. Francis Lefebure (September 17, 1916 – March 19, 1988) was a French physician, known for his study of phosphenes.

Early Life

Lefebure was born in Paris, France. His mother, Claire de Saint-Rémy, was a painter and poet. His father, a lawyer, published a magazine about thermal and tourist laws. He attended the Lycée Louis-le-Grand, the high school which has produced the largest number of famous French men. All his life, Dr. Lefebure kept a fond memory of his science teachers. At age 17, he attended the "P.C.N." (Physics, Chemistry, Biology), which was meant as a preparatory year before the study of Medicine. He graduated among the first of several hundred students, although he was the youngest.

He started at the age of 13, simultaneously to his academic studies, to express an interest in methods of personal development, yoga and spiritual experiences. This was mainly because Dr. Lefebure lived in the famous 'Latin quarter' of Paris, which was the only place where bookshops specialised in these fields could be found at the time. His early research was greatly supported by his mother, who knew about her son's interest and who would find him the best books. At the early age of 15, he had already passionately studied all the works of the philosopher and founder of anthroposophy, Rudolf Steiner. Though much later, after his discoveries on the phosphenes, Dr. Lefebure distanced himself from Steiner's work.

At the age of 18, he began to study medicine. His main interests lied in anatomy, histology, physiology and physics applied to medicine, fields in which he obtained top marks but, like Darwin, he never had any interest in pathology or clinical medicine, though he had to study those subjects anyway. He became an extern of the Hospitals of Paris, choosing the neurology and children psychiatry services.

Meeting Galip, and Army Life

In June 1934, he met the man who in a way became his mentor: Artheme Galip, a Ukrainian diplomat, son of the last governor of Bucovinia, before it was annexed by the U.S.S.R. After the annexation, Galip moved to France as a refugee. He had previously spent time in a zoroastrian temple. This man was supposed to have the ability to trigger visions by laying on of hands, but had no proper teaching. There was a contradiction between his gift and the exercises he taught, some of them were actually very harmful, others were just a waste of time.

Galip left for South America and disappeared. Dr Lefebure, while he continued his medical studies, was left alone trying to untangle the contradictory effects of the laying on of hands and of the exercises Galip had taught him. For that purpose, he passed his doctorate in 1942 in Algiers with a thesis on the breathing exercises of Yoga: "Rhythmic breathing and mental concentration". He was drafted in 1939 as an auxiliary medic and was later promoted to the rank of lieutenant medic.

He used the long years of draft to ponder all that he had previously learnt. From his meditations in the army came a book: "Homologies, the Analogies of the Microcosm and Macrocosm", which he considered as his best work on the intellectual level. It is basically an extension of the theory of symmetry, a path that academic science would follow, more than 40 years later, with the studies on "Fractals" and "Internal Homothecy".

Biological Studies

After leaving the army, Dr. Lefebure decided to follow his lifetime vocation and started a degree in biology, then taught science. But, being married, he had to accept a post as a school hygiene doctor. He nevertheless continued his research and created the association 'Homosophy' (the Wisdom of Similarities), thus creating the first groups of study of initiatory techniques, in 1945.

While bringing up three children, he spent his spare time writing another book, "The Power of Christianity", to prove that Christianity was built on exercises that were later taken out and also to demonstrate that non-violence is the best way of defence. This was the time when he wrote about the effects Galip's laying on of hands had triggered in his book "Initiatory Experiences" and about the out-of-body experiences that his method (published in "Homologies") had produced.

Meeting with Subuh

In 1959, his meeting with the Indonesian mage Pak Subuh was decisive, as Dr Lefebure wanted to compare Subuh's exercises with those of his first master. Electroencephalograms were impossible to use as they require muscular relaxation, so he had the idea to monitor the action on a phosphene of the different variations of a same exercise. This is how he discovered the first phosphenic phenomenon that had never been reported before: if the head is moved on a rhythm of two seconds, the phosphene follows that movement. If the movement of the head is faster or slower, the phosphene stays fixed.

This led him to experimenting with what would happen with two phosphenes. The method of brain exploration by the study of the rhythms of double phosphenes seemed to be of extraordinary richness and practical use to him. He wrote a book on the subject. Its first draft was a report that he transmitted to the School Health Service in the spring of 1960. This book describes various rhythmic and cerebral phenomena which were unknown before and that became possible to observe thanks to the Cerebroscope, a device for which the Lefebure was awarded the Silver Medal at the Lépine contest of inventors in 1964. He then applied this discovery to the ear and designed the Synchrophone which later became the alternophone, a cerebral activation device and wrote "Activating the brain with alternative hearing". He was awarded the Gold Medal and first prize of the Lépine contest in 1963 and the Gold Medal at the International Inventors Fair in Brussels in 1964 for the action on the brain of the alternative hearing device (Alternophonia).

Phosphenic Mixing

In 1963, he had the idea to study what happens when an individual thinks about a particular subject during the presence of the phosphene. This idea was the origin of the principle of "Phosphenic Mixing", a method that supposedly develops memory, intelligence, the sense of initiative, artistic intuition and creativity by mixing thoughts and phosphenes. In November 1975, the World's Inventors Fair in Brussels awarded Lefebure the Silver Medal for his method: "Phosphenic Mixing Applied to Education" which in 2004, has been printed 16 times and has sold more than 58 thousand copies.

He thought that the method was fruitful and remarkably easy to apply. Since then, he spent all his energy on perfecting this technique, teaching it and applying it to other sensory organs.

Every reader that would relate their application of the method to him became collaborators (their names have always been quoted alongside the fact that have discovered, unless Dr Lefebure was asked not to). Thus, after twenty nine years of work, he managed to accumulate a considerable and unique documentation.

The collected information underscored the part that Phosphenism could have played in the origin of religions and lead to the publication of a book: "Phosphenism and the Origin of Religions" ("Developing Clairvoyance by using Phosphenism").

It is during this period that he thought he solved the great enigma of his youth: where did Galip's strange power of laying on of hands came from? Indeed, he believed that phosphenes could be very easily transmissible by telepathy. The zoroastrian areas were the only places in the pre-industrial world where oil naturally sprang from the ground. Thus, since the dawn of prehistory, a stock of human beings has had a much greater material ease for producing bright and permanent fires, though they could not analyse the details of the process: thinking while focusing on a bright fire. Without being aware of it, they practiced Phosphenism. This could be why in this area, there are individuals who are supposed to possess a secret knowledge, that could be magically explained because they have particular gifts produced by the instinctive practice of Phosphenic Mixing.

Later, the extension of the principle of Phosphenic Mixing to breathing led him to write "The pneumophene, or the breathing techniques that open the doors to the world beyond". A similar extension to the sense of movement led Lefebure to write "From the Praying Wheel to the

Spiritual Dynamo, or the Kundalini-triggering device". This book, based on supposed scientific experiments, shines a new light on eastern esoteric traditions. It was followed in 1988 by volume II that describes new applications of the Gyrascope. Lefebure considered that "thanks to the phosphene and the various types of Gyrascopes, a door has been opened in 'esoterism'. The phenomena qualified of initiatory are finally open to everyone. The force liberated is like burning lava sweeping humanity."

Meeting Daniel Stiennon was a deciding factor for the diffusion of phosphenism. He created an association then, very quickly a company that is now present in different points of the world.

The works of Lefebure can be compared to the works of the greatest esotericists. After thirty years of organising lectures and seminars all over the world, and several of the books having been translated in various languages, some think that Lefebure has had an influence on researchers since 1945, but many others don't. Mrs Alexandra David-Neel had, for example, studied the book that is the cornerstone of Phosphenism: "Homologies" and knew Lefebure. Of course, many drew influence from his works without ever quoting him. On the contrary, Lefebure always quoted his sources.

A Few Dates in the History of Phosphenism

- 1916: Dr. Francis Lefebure was born on September 17, 1916, in Paris, France. He died on March 19, 1988 in his apartment 104 rue Réaumur Paris, France.
- 1945: He created the association "Homosophy - Wisdom of Similarities", which shortly afterwards became the association "Phosphenism".
- 1959: The first discovery of Dr. Lefebure about phosphenes, and the release of his first book on Phosphenism.
- 1960: Verification of the discoveries of Dr. Lefebure by the C.N.R.S, at the National Institute of Sports Laboratory and the Henri Roussel Hospital, Paris.
- 1963: First Gold medal and Prize at the "Lépine" contest.
- 1967: Dispatching of telegrams and letters to the Vatican and to his Holiness, Paul VI, revealing the scientific explanation of the "phenomena" that took place in Fatima, Portugal. After these revelations that "disturbed" the pope, the Church gave up recognising "solar prodigies" as specifically catholic miracles.

This is an implicit acknowledgement of the discoveries of Dr Lefebure.

- 1967: First photographs of phosphenes, realised by Louis Cuttica. Over a hundred of such photographs were supposedly taken.
- 1969: The secret services of several countries, including France, took an interest in Dr. Lefebure's discoveries. However, they closed the matter quickly.
- 1974: Publication of "Phosphenism, a new explanation of the origin of religions".
- 1975-1982: Series of conferences in Germany, England, Portugal, the Island of Réunion, Quebec (Canada) and in parts of Africa.
- 1981: First meeting of Daniel Stiennon and Dr. Lefebure.
- 1981: After describing something called "rotational function of the brain", Dr. Lefebure created the "Gyrascope".
- 1986: Creation of the company "Editions Phosphénisme" by Daniel Stiennon.
- 1996: "Phosphenic Mixing Applied to Education" reaches its 14th edition and has sold over 56.000 copies.
- 1997: Creation of several web sites on the Internet in three languages (French, Spanish and English).
- 1999: Opening of the Phosphenology Research Centre in the French department of Gard.
- 2001: E-books editions of Dr. Lefebure's books available on the Internet.

Achievements and Awards

- Former Extern of the Hospitals of Paris, France
- Former Physician at the School Health Services
- Former teacher of science and mathematics.
- Gold medal and prize at the "Lépine" contest, 1963
- Silver medal at the "Lépine" contest, 1964
- Gold medal at the International Inventor's Fair, 1964
- Vermilion medal at the International Inventor's Fair, 1967
- Silver medal at the International Inventor's Fair, Brussels, 1975, for his method "Phosphenic Mixing Applied to Education".

Purkinje Images

Purkinje images are reflections of objects from structure of the eye. They are also known as Purkinje reflexes and as Purkinje-Sanson images. There are at least four Purkinje images that are visible on looking at an eye. The first Purkinje image (P1) is the reflection from the outer surface of the cornea. The second Purkinje image (P2) is the reflection from the inner surface of the cornea. The third Purkinje image (P3) is the reflection from the outer (anterior) surface of the lens. The fourth Purkinje image (P4) is the reflection from the inner (posterior) surface of the lens. Unlike the others, P4 is an inverted image. Purkinje-Sanson images are named after Czech anatomist Jan Evangelista Purkynµ (1787-1869) and after French physician Louis Joseph Sanson (1790-1841).

The third and fourth Purkinje images can be visible from within the eye itself. Light reflected away from the surfaces of the lens can in turn reflect back into the eye from the rear surface of the cornea. These images are, therefore, entoptic phenomena.

The first and fourth Purkinje images are used by some eye trackers, devices to measure the position of an eye.

Brightness of Purkinje Images

The brightness of the Purkinje images can be determined using Fresnel's equation where the Intensity (Brightness) = $(n' " n)^2 /(n' + n)^2$ Purkinje image I is the brightest of the four then PII, PIII and PIV (III and IV have about the same brightness). The Purkinje images can be used to assess the curvatures and separations of the surfaces in the eye.

Red-eye Effect

The red-eye effect in photography is the common appearance of red pupils in colour photographs of eyes. It occurs when using a photographic flash very close to the camera lens (as with most compact cameras), in ambient low light. The effect appears in the eyes of humans and animals that have no tapetum lucidum, hence noeyeshine, and rarely in animals that have a tapetum lucidum. The red-eye effect can also be seen when one human observes another when there is a bright lightsource nearby.

Theatrical followspot operators, positioned nearly coincidentally with a very bright light and somewhat distant from the actors, occasionally witness red-eye in actors on stage. The effect is not visible to the rest of the audience because it is reliant on the very small angle between the followspot operator and the light. The feature film Blade Runner intentionally induced a red-eye effect to indicate that certain characters were "replicants" (according to the story, an artificial life-form). The effect relied on a semi-silvered mirror directing light at the actors on the same incidence as the camera.

Causes

Because the light of the flash occurs too fast for the pupil to close, much of the very bright light from the flash passes into the eye through the pupil, reflects off the fundus at the back of the eyeball and out through the pupil. The camera records this reflected light. The main cause of the red colour is the ample amount of blood in the choroid which nourishes the back of the eye and is located behind the retina. The blood in the retinal circulation is far less than in the choroid, and plays virtually no role.

The eye contains several photostable pigments that all absorb in the short wavelength region, and hence contribute somewhat to the red eye effect. Thelens cuts off deep blue and violet light, below 430 nm (depending on age), and macular pigment absorbs between 400 and 500 nm, but this pigment is located exclusively in the tiny fovea. Melanin, located in the retinal pigment epithelium (RPE) and the choroid, shows a gradually increasing absorption towards the short wavelengths.

But blood is the main determinant of the red colour, because it is completely transparent at long wavelengths and abruptly starts absorbing at 600 nm. The amount of red light emerging from the pupil depends on the amount of melanin in the layers behind the retina. This amount varies strongly between individuals. Light skinned people with blue eyes have relatively low melanin in the fundus and thus show a much stronger red-eye effect than dark skinned people with brown eyes.

The same holds for animals. The colour of the iris itself is of virtually no importance for the red-eye effect. This is obvious because the red-eye effect works best when photographing dark adapted subjects, hence with fully dilated pupils.

Photographs taken with infrared light through night vision devices always show very bright pupils because, in the dark, the pupils are fully dilated and the infrared light is not absorbed by any ocular pigment. The role of melanin in red-eye effect is nicely demonstrated in animals with heterochromia: only the blue eye displays the effect. The effect is still more pronounced in humans and animals with albinism. All forms of albinism involve abnormal production and/or deposition of melanin.

Red-eye effect is seen in photographs of children also because children's eyes have more rapid dark adaption: in low light a child's pupils enlarge sooner, and an enlarged pupil accentuates the red-eye effect.

In a photograph of a child's face, red-eye in one eye but not the other may be a sign of the cancerretinoblastoma.

Similar Effects

Similar effects, some related to red-eye effect, are of several kinds:

- In many flash photographs, even those without perceptible red-eye effect, many animals' pupils displayeyeshine. Although eyeshine is an unrelated effect, animals with blue eyes may display the red-eye effect in addition to eyeshine.
- Numerous diseases of the eye cause a "white-eye effect" instead of the expected red-eye effect.
- A related effect, red reflex, is seen in fundoscopy; unlike the red-eye effect, it is not a photographic effect.
- In photographs recorded with infrared-sensitive passive (non-IR emitting) equipment, the eyes (not only the pupils) usually appear very bright. This is due not to reflection, but to radiation of core body heat in the form of infrared light.

Photography Techniques for Prevention and Removal

The red-eye effect can be prevented in a number of ways.

1. Using bounce flash in which the flash head is aimed at a nearby pale coloured surface such as a ceiling or wall or at a specialist photographic reflector. This both changes the direction of the flash and ensures that only diffused flash light enters the eye.
2. Placing the flash away from the camera's optical axis ensures that the light from the flash hits the eye at an oblique angle. The light enters the eye in a direction away from the optical axis of the camera and is refocused by the eye lens back along the same axis. Because of this the retina will not be visible to the camera and the eyes will appear natural.

3. Taking pictures without flash by increasing the ambient lighting, opening the lens aperture, using a faster film or detector, or reducing the shutter speed.
4. Using the red-eye reduction capabilities built into many modern cameras. These precede the main flash with a series of short, low-power flashes, or a continuous piercing bright light triggering the pupil to contract. (This should not be confused with the autofocus assist beam, which uses a series of flashes for focus instead.)
5. Having the subject look away from the camera lens.
6. The red-eye effect is not seen in subjects wearing contact lenses with UV filtering
7. Increase the lighting in the room so that the subject's pupils are more constricted.
8. Some computer digital image editors have the ability to lessen the red eye by desaturating the red colour.

If direct flash must be used, a good rule of thumb is to separate the flash from the lens by 1/20 of the distance of the camera to the subject. For example, if the subject is 2 metres (6 feet) away, the flash head should be at least 10 cm (4 inches) away from the lens. Professional photographers prefer to use ambient light or indirect flash as the red-eye reduction system does not always prevent red eyes, for example if people look away during the pre-flash. In addition, people do not look natural with small pupils, and direct lighting from close to the camera lens is considered to produce unflattering photographs. Various graphics editing software packages have functions to remove red eyes from digital photographs. Some have automated this process, and can apply it to many photos at once. Red-eye removal is also built into many popular consumer programs, such as Adobe Photoshop, Apple iPhoto, Corel Photo-Paint, GIMP, Google Picasa, and Microsoft Windows Live Photo Gallery.

Tapetum Lucidum

The *tapetum lucidum* (Latin: "bright tapestry", plural tapeta *lucida*) is a layer of tissue in the eye of many vertebrate animals. It lies immediately behind or sometimes within the retina. Itreflects visible light back through the retina, increasing the light available to the photoreceptors. This improves vision in low-light conditions, but can cause the perceived image to be blurry from the interference of the reflected light. The tapetum lucidum contributes to the superior night vision of some animals. Many of these animals are nocturnal, especially carnivores that hunt their prey at night, while others are deep sea animals.

Eyeshine

Eyeshine is a visible effect of the *tapetum lucidum*. When light shines into the eye of an animal having a *tapetum lucidum*, the pupil appears to glow. Eyeshine can be seen in many animals, in nature and in flash photographs. In low light, a hand-held flashlight is sufficient to produce eyeshine that is highly visible to humans (despite our inferior night vision); this technique, spotlighting, is used by naturalists and hunters to search for animals at night. Eyeshine occurs in a wide variety of colours including white, blue, green, yellow, pink and red. However, because eyeshine is a form of iridescence, the colour varies slightly with the angle at which it is seen and the colour of the source light.

Figure: In darkness, eyeshine reveals this raccoon

White eyeshine occurs in many fish, especially walleye; blue eyeshine occurs in many mammals such as horses; yellow eyeshine occurs in mammals such as cats, dogs, and raccoons; and red eyeshine occurs inrodents, opossums and birds. The human eye has no *tapetum lucidum*, hence no eyeshine. However, in humans and animals two effects can occur that may resemble eyeshine: *leukocoria* (white shine, indicative of abnormalities including cataracts, cancers, and other problems) and red-eye effect.

Blue-eyed Cats and Dogs

Cats and dogs with blue eyes may display both eyeshine and red-eye effect. Both species have a tapetum lucidum, so their pupils may display eyeshine. In flash colour photographs, however, individuals with blue eyes may also display a distinctive red eyeshine. Individuals with heterochromia may display red eyeshine in the blue eye and "normal" yellow/green/blue/white eyeshine in the other eye. These include odd-eyed cats and bi-eyed dogs. The red-eye effect is independent of the eyeshine: in some photographs of individuals with a tapetum lucidum and heterochromia, the eyeshine is dim yet the pupil of the blue eye still appears red. This is most apparent when the individual is not looking into the camera, because the tapetum lucidum is far less extensive than the retina.

Classification

A classification of anatomical variants of tapeta lucida defines 4 types:

- *Retinal tapetum*, as seen in teleosts, crocodiles, marsupials and fruit bats. The tapetum lucidum is within the retina; in the other 3 types the tapetum is within the choroid behind the retina.
- *Choroidal guanine tapetum*, as seen in elasmobranchii (skates, rays, and sharks) and chimaeras. The tapetum is a palisade of cells containing stacks of flat hexagonal crystals of guanine.
- *Choroidal tapetum cellulosum*, as seen in carnivores, rodents and cetacea. The tapetum consists of layers of cells containing organised, highly refractive crystals. These crystals are diverse in shape and makeup.
- *Choroidal tapetum fibrosum*, as seen in cows, sheep, goats and horses. The tapetum is an array of extracellular fibres.

The functional differences between these 4 different types of tapeta lucida are not known. This classification does not include tapeta lucida in birds. Kiwis, Stone-curlews, the Boat-billed Heron, the flightless Kakapo and many Nightjars, Owls, and other night birds such as the Swallow-tailed Gullalso possess a tapetum lucidum This classification also does not include the extraordinary focusing mirror in the eye of the brownsnout spookfish.

Like humans, some animals lack a tapetum lucidum and they usually are diurnal. These include most primates, squirrels, some birds, red kangaroo, and pig. Primates that have a tapetum lucidum include the aye and sportive lemur. When a *tapetum lucidum* is present, its location on the eyeball varies with the placement of the eyeball in the head, such that in all cases the tapetum lucidum enhances night vision in the centre of the animal's field of view. Apart from its eyeshine, the *tapetum luminum* itself has a colour. It is often described as iridescent.

Intigers it is greenish. In ruminants it may be golden green with a blue periphery, or whitish or pale blue with a lavender periphery. In dogs it may be whitish with a blue periphery.

Mechanism

The *tapetum lucidum*, which is iridescent, reflects light roughly on the interference principles of thin-film optics, as seen in other iridescent tissues. However, the *tapetum lucidum* cells are leucophores, notiridophores. The *tapetum lucidum* is a retroreflector of the transparent sphere type. Because it is a retroreflector, it reflects light directly back along the light path.

This serves to match the original and reflected light, thus maintaining the sharpness and contrast of the image on the retina. The tapetum *lucidum* reflects with constructive interference, thus increasing the quantity of light passing through the retina.

In the cat, the *tapetum lucidum* lowers the minimum threshold of vision 6-fold, allowing the cat to see light that is invisible to human eyes.

Uses

The *tapetum lucidum* enables an animal with it to see in dimmer light than the animal would otherwise be able to. This is of use to the animal, but it is of use also to humans.

Human uses include scanning for reflected eyeshine to detect and identify the species of animals in the dark, and deploying trained search dogs and search horses at night. Historically, its function was regarded as simply to increase the light intensity of an image on the retina.

Using eyeshine to identify animals in the dark employs not only its colour but also several other features. The colour corresponds approximately to the type of tapetum lucidum, with some variation between species.

Other features include the distance between pupils relative to their size; the height above ground; the manner of blinking (if any); and the movement of the eyeshine (bobbing, weaving, hopping, leaping, climbing, flying).

Manufactured retroreflectors modelled after a tapetum lucidum are described in numerous patents and today have many uses. The earliest patent, first used in "Catseye" brand raised pavement markers, was inspired by the tapetum lucidum of a cat's eye. A more recent use of retroreflectors, helping to provide secure communications between two stations in line of sight, is modelled after the combination of tapetum lucidum and bioluminescent "flashlight" in flashlight fish of the families Anomalopidae andStomiidae.

Some of these fish have been shown to use eyeshine both to detect and to communicate with other flashlight fish.

Pathology

In dogs, certain drugs are known to disturb the precise organisation of the crystals of the tapetum lucidum, thus compromising the dog's ability to see in low light.

These drugs include ethambutol, macrolide antibiotics, dithizone, antimalarial compounds, some receptor H(sub)2-antagonists, and cardiovascular agents. The disturbance "is attributed to the chelating action which removes zinc from the tapetal cells."

2

Waves in Encyclopaedia

Speed of Propagation

Any electric charge that accelerates, or any changing magnetic field, produces electromagnetic radiation. Electromagnetic information about the charge travels at the speed of light.

Accurate treatment thus incorporates a concept known as retarded time (as opposed to advanced time, which is not physically possible in light of causality), which adds to the expressions for the electrodynamic electric field and magnetic field.

These extra terms are responsible for electromagnetic radiation. When any wire (or other conducting object such as an antenna) conducts alternating current, electromagnetic radiation is propagated at the same frequency as the electric current.

At the quantum level, electromagnetic radiation is produced when the wavepacket of a charged particle oscillates or otherwise accelerates. Charged particles in a stationary state do not move, but a superposition of such states may result in oscillation, which is responsible for the phenomenon of radiative transition between quantum states of a charged particle.

Depending on the circumstances, electromagnetic radiation may behave as a wave or as particles. As a wave, it is characterised by a velocity (the speed of light), wavelength, and frequency. When considered as particles, they are known as photons, and each has an energy related to the frequency of the wave given by Planck's relation $E = h\nu$, where E is the energy of the photon, $h = 6.626 \times 10^{"34}$J ·s is Planck's constant, and ν is the frequency of the wave.

One rule is always obeyed regardless of the circumstances: EM radiation in a vacuum always travels at the speed of light, *relative to the observer*, regardless of the observer's velocity. (This observation led to Albert Einstein's development of the theory of special relativity.) In a medium (other than vacuum), velocity factor or refractive index are considered, depending on frequency and application. Both of these are ratios of the speed in a medium to speed in a vacuum.

Thermal Radiation and Electromagnetic Radiation as a form of Heat

The basic structure of matter involves charged particles bound together in many different ways. When electromagnetic radiation is incident on matter, it causes the charged particles to oscillate and gain energy. The ultimate fate of this energy depends on the situation. It could be immediately re-radiated and appear as scattered, reflected, or transmitted radiation. It may also get dissipated into other microscopic motions within the matter, coming to thermal equilibrium and manifesting itself as thermal energy in the material. With a few exceptions (such as fluorescence, harmonic generation, photochemical reactions and the photovoltaic effect), absorbed electromagnetic radiation simply deposits its energy by heating the material. This happens both for infrared and for non-infrared radiation. Intense radio waves can thermally burn living tissue and can cook food. In addition to infrared lasers, sufficiently intense visible and ultraviolet lasers can also easily set paper afire. Ionizing electromagnetic radiation can create high-speed electrons in a material and break chemical bonds, but after these electrons collide many times with other atoms in the material eventually most of the energy gets downgraded to thermal energy; this whole process happens in a tiny fraction of a second. It is a widespread misconception in physics that infrared radiation is a form of heat and that other electromagnetic radiation is not; *any* electromagnetic radiation can heat a material when it is absorbed.

The inverse or time-reversed process of absorption is responsible for thermal radiation. Much of the thermal energy in matter consists of random motion of charged particles, and this energy can be radiated away from the matter. The resulting radiation may subsequently be absorbed by another piece of matter, with the deposited energy heating the material. Radiation is an important mechanism of heat transfer.

The electromagnetic radiation in an opaque cavity at thermal equilibrium is effectively a form of thermal energy, having maximum radiation entropy. The thermodynamic potentials of electromagnetic radiation can be well-defined as for matter. Thermal radiation in a cavity has energy density of

$$\frac{U}{V}=\frac{8\pi^5(kT)^4}{15(hc)^3},$$

Differentiating the above with respect to temperature, we may say that the electromagnetic radiation field has an effective volumetric heat capacity given by

$$\frac{32\pi^5k^4T^3}{15(hc)^3},$$

Electromagnetic Spectrum

In general, EM radiation (the designation 'radiation' excludes static electric and magnetic and near fields) is classified by wavelength into radio, microwave, infrared, the visible region we perceive as light, ultraviolet, X-rays, and gamma rays. Arbitrary electromagnetic waves can always be expressed by Fourier analysis in terms of sinusoidal monochromatic waves, which can be classified into these regions of the spectrum.

The behaviour of EM radiation depends on its wavelength. Higher frequencies have shorter wavelengths, and lower frequencies have longer wavelengths. When EM radiation interacts with single atoms and molecules, its behaviour depends on the amount of energy per quantum it carries. Spectroscopy can detect a much wider region of the EM spectrum than the visible range of 400 nm to 700 nm. A common laboratory spectroscope can detect wavelengths from 2 nm to 2500 nm. Detailed information about the physical properties of objects, gases, or even stars can be obtained from this type of device. It is widely used in astrophysics. For example, hydrogen atoms emit radio waves of wavelength 21.12 cm.

Soundwaves are not electromagnetic radiation. At the lower end of the electromagnetic spectrum, about 20 Hz to about 20 kHz, are frequencies that might be considered in the audio range. However, electromagnetic waves cannot be directly perceived by human ears. Sound waves are the oscillating compression of molecules. To be heard, electromagnetic radiation must be converted to pressure waves of the fluid in which the ear is located (whether the fluid is air, water or something else).

Light

EM radiation with a wavelength between approximately 400 nm and 700 nm is directly detected by the human eye and perceived as visiblelight. Other wavelengths, especially nearby infrared (longer than 700 nm) and ultraviolet (shorter than 400 nm) are also sometimes referred to as light, especially when visibility to humans is not relevant. If radiation having a frequency in the visible region of the EM spectrum reflects off of an object, say, a bowl of fruit, and then strikes our eyes, this results in our visual perception of the scene.

Our brain's visual system processes the multitude of reflected frequencies into different shades and hues, and through this not-entirely-understood psychophysical phenomenon, most people perceive a bowl of fruit.

At most wavelengths, however, the information carried by electromagnetic radiation is not directly detected by human senses. Natural sources produce EM radiation across the spectrum, and our technology can also manipulate a broad range of wavelengths. Optical fibre transmits light, which, although not suitable for direct viewing, can carry data that can be translated into sound or an image. To be meaningful both transmitter and receiver must use some agreed-upon encoding system - especially so if the transmission is digital as opposed to the analogue nature of the waves.

Radio Waves

Radio waves can be made to carry information by varying the amplitude, frequency or phase.

When EM radiation impinges upon a conductor, it couples to the conductor, travels along it, and induces an electric current on the surface of that conductor by exciting the electrons of the conducting material. This effect (the skin effect) is used in antennas. EM radiation may also cause certain molecules to absorb energy and thus to heat up; this is exploited in microwave ovens. Radio waves are *not* ionizing radiation, as the energy per photon is too small.

Derivation

Electromagnetic waves as a general phenomenon were predicted by the classical laws of electricity and magnetism, known as Maxwell's equations. Inspection of Maxwell's equations without sources (charges or currents) results in, along with the possibility of nothing happening, nontrivial solutions of changing electric and magnetic fields. Beginning with Maxwell's equations in free space:

$$\nabla \cdot \mathbf{E} = 0 \qquad (1)$$

$$\nabla \times \mathbf{E} = -\frac{\partial \mathbf{B}}{\partial t} \qquad (2)$$

$$\nabla \cdot \mathbf{B} = 0 \qquad (3)$$

$$\nabla \times \mathbf{B} = \mu_0 \epsilon_0 \frac{\partial \mathbf{E}}{\partial t} \qquad (4)$$

where

∇ is a vector differential operator.

One solution,

$$\mathbf{E} = \mathbf{B} = \mathbf{0},$$

is trivial. For a more useful solution, we utilise vector identities, which work for any vector, as follows:

$$\nabla \times (\nabla \times \mathbf{A}) = \nabla (\nabla \cdot \mathbf{A}) - \nabla^2 \mathbf{A}$$

To see how we can use this, take the curl of equation (2):

$$\nabla \times (\nabla \times \mathbf{E}) = \nabla \times \left(-\frac{\partial \mathbf{B}}{\partial t} \right) \qquad (5)$$

Evaluating the left hand side:

$$\nabla \times (\nabla \times \mathbf{E}) = \nabla (\nabla \cdot \mathbf{E}) - \nabla^2 \mathbf{E} = -\nabla^2 \mathbf{E} \qquad (6)$$

where we simplified the above by using equation (1).

Evaluate the right hand side:

$$\nabla \times \left(-\frac{\partial \mathbf{B}}{\partial t} \right) = -\frac{\partial}{\partial t} (\nabla \times \mathbf{B}) = -\mu_0 \epsilon_0 \frac{\partial^2 \mathbf{E}}{\partial t^2} \qquad (7)$$

Equations (6) and (7) are equal, so this results in a vector-valued differential equation for the electric field, namely

$$\nabla^2 \mathbf{E} = \mu_0 \epsilon_0 \frac{\partial^2 \mathbf{E}}{\partial t^2}$$

Applying a similar pattern results in similar differential equation for the magnetic field:

$$\nabla^2 \mathbf{B} = \mu_0 \epsilon_0 \frac{\partial^2 \mathbf{B}}{\partial t^2}.$$

These differential equations are equivalent to the wave equation:

$$\nabla^2 f = \frac{1}{c_0^2} \frac{\partial^2 f}{\partial t^2}$$

where

c_0 is the speed of the wave in free space and

f describes a displacement

Or more simply:

$$\Box f = 0$$

where $\Box$ is d'Alembertian:

$$\Box = \nabla^2 - \frac{1}{c_0{}^2}\frac{\partial^2}{\partial t^2} = \frac{\partial^2}{\partial x^2} + \frac{\partial^2}{\partial y^2} + \frac{\partial^2}{\partial z^2} - \frac{1}{c_0{}^2}\frac{\partial^2}{\partial t^2}$$

Notice that, in the case of the electric and magnetic fields, the speed is:

$$c_0 = \frac{1}{\sqrt{\mu_0 \epsilon_0}}$$

This is the speed of light in vacuum. Maxwell's equations have unified the vacuum permittivity ϵ_0, the vacuum permeability μ_0, and the speed of light itself, c_0. Before this derivation it was not known that there was such a strong relationship between light and electricity and magnetism. But these are only two equations and we started with four, so there is still more information pertaining to these waves hidden within Maxwell's equations. Let's consider a generic vector wave for the electric field.

$$\mathbf{E} = \mathbf{E}_0 f\left(\hat{\mathbf{k}}\cdot\mathbf{x} - c_0 t\right)$$

Here, $\mathbf{E}_0$ is the constant amplitude, f is any second differentiable function, $\hat{\mathbf{k}}$ is a unit vector in the direction of propagation, and $\mathbf{x}$ is a position vector. We observe that $f\left(\hat{\mathbf{k}}\cdot\mathbf{x} - c_0 t\right)$ is a generic solution to the wave equation. In other words

$$\nabla^2 f\left(\hat{\mathbf{k}}\cdot\mathbf{x} - c_0 t\right) = \frac{1}{c_0{}^2}\frac{\partial^2}{\partial t^2} f\left(\hat{\mathbf{k}}\cdot\mathbf{x} - c_0 t\right),$$

for a generic wave travelling in the $\hat{\mathbf{k}}$ direction.

This form will satisfy the wave equation, but will it satisfy all of Maxwell's equations, and with what corresponding magnetic field?

$$\nabla\cdot\mathbf{E} = \hat{\mathbf{k}}\cdot\mathbf{E}_0 f'\left(\hat{\mathbf{k}}\cdot\mathbf{x} - c_0 t\right) = 0$$

$$\mathbf{E}\cdot\hat{\mathbf{k}} = 0$$

The first of Maxwell's equations implies that electric field is orthogonal to the direction the wave propagates.

$$\nabla \times \mathbf{E} = \hat{\mathbf{k}} \times \mathbf{E}_0 f'\left(\hat{\mathbf{k}} \cdot \mathbf{x} - c_0 t\right) = -\frac{\partial \mathbf{B}}{\partial t}$$

$$\mathbf{B} = \frac{1}{c_0} \hat{\mathbf{k}} \times \mathbf{E}$$

The second of Maxwell's equations yields the magnetic field. The remaining equations will be satisfied by this choice of $\mathbf{E}, \mathbf{B}$.

Not only are the electric and magnetic field waves travelling at the speed of light but they have a special restricted orientation and proportional magnitudes, $E_0 = c_0 B_0$, which can be seen immediately from the Poynting vector. The electric field, magnetic field, and direction of wave propagation are all orthogonal, and the wave propagates in the same direction as $\mathbf{E} \times \mathbf{B}$.

From the viewpoint of an electromagnetic wave travelling forward, the electric field might be oscillating up and down, while the magnetic field oscillates right and left; but this picture can be rotated with the electric field oscillating right and left and the magnetic field oscillating down and up. This is a different solution that is travelling in the same direction. This arbitrariness in the orientation with respect to propagation direction is known as polarisation. On a quantum level, it is described as photon polarisation. The direction of the polarisation is defined as the direction of the electric field.

More general forms of the second-order wave equations given above are available, allowing for both non-vacuum propagation media and sources. A great many competing derivations exist, all with varying levels of approximation and intended applications. One very general example is a form of the electric field equation, which was factorized into a pair of explicitly directional wave equations, and then efficiently reduced into a single unidirectional wave equation by means of a simple slow-evolution approximation.

Wave

In physics, a wave is a disturbance (an oscillation) that travels through space and time, accompanied by the transfer of energy.

Waves travel and the wave motion transfers energy from one point to another, often with no permanent displacement of the particles of the medium—that is, with little or no associated mass transport. They consist, instead, of oscillations or vibrations around almost fixed locations. For example, a cork on rippling water will bob up and down, staying in about the same place while the wave itself moves onwards.

One type of wave is a mechanical wave, which propagates through a medium in which the substance of this medium is deformed. The deformation reverses itself owing to restoring forces resulting from its deformation. For example, sound waves propagate via air molecules bumping into their neighbours. This transfers some energy to these neighbours, which will cause a cascade of collisions between neighbouring molecules. When air molecules collide with their neighbours, they also bounce away from them (restoring force). This keeps the molecules from continuing to travel in the direction of the wave.

Another type of wave can travel through a vacuum, e.g. electromagnetic radiation (including visible light, ultraviolet radiation, infrared radiation, gamma rays, X-rays, and radio waves). This type of wave consists of periodic oscillations in electrical and magnetic fields.

A main distinction can be made between transverse and longitudinal waves. Transverse waves occur when a disturbance creates oscillations perpendicular (at right angles) to the propagation (the direction of energy transfer). Longitudinal waves occur when the medium oscillates in the same direction as the propagation.

Waves are described by a wave equation which sets out how the disturbance proceeds over time. The mathematical form of this equation varies depending on the type of wave.

General Features

A single, all-encompassing definition for the term *wave* is not straightforward. A vibration can be defined as a *back-and-forth* motion around a reference value. However, a vibration is not necessarily a wave. An attempt to define the necessary and sufficient characteristics that qualify a phenomenon to be called a *wave* results in a fuzzy border line.

The term *wave* is often intuitively understood as referring to a transport of spatial disturbances that are generally not accompanied by a motion of the medium occupying this space as a whole. In a wave, the energy of a vibration is moving away from the source in the form of a disturbance within the surrounding medium (Hall 1980, p. 8). However, this notion is problematic for a standing wave (for example, a wave on a string), where energy is moving in both directions equally, or for electromagnetic/light waves in a vacuum, where the concept of medium does not apply and the inherent interaction of its component is the main reason of it's motion and broadcasting. There are water waves on the ocean surface; light waves emitted by the Sun; microwaves used in microwave ovens; radio waves broadcast by radio stations; and sound waves generated by radio receivers, telephone handsets and living creatures (as voices).

It may appear that the description of waves is closely related to their physical origin for each specific instance of a wave process. For example, acoustics is distinguished from optics in that sound waves are related to a mechanical rather than an electromagnetic wave transfer caused by vibration.

Concepts such as mass, momentum, inertia, or elasticity, become therefore crucial in describing acoustic (as distinct from optic) wave processes. This difference in origin introduces certain wave characteristics particular to the properties of the medium involved. For example, in the case of air:vortices, radiation pressure, shock waves etc.; in the case of solids: Rayleigh waves, dispersion etc.; and so on.

Other properties, however, although they are usually described in an origin-specific manner, may be generalised to all waves. For such reasons, wave theory represents a particular branch of physics that is concerned with the properties of wave processes independently from their physical origin. For example, based on the mechanical origin of acoustic waves, a moving disturbance in space–time can exist if and only if the medium involved is neither infinitely stiff nor infinitely pliable. If all the parts making up a medium were rigidly *bound*, then they would all vibrate as one, with no delay in the transmission of the vibration and therefore no wave motion. This is impossible because it would violate general relativity. On the other hand, if all the parts were independent, then there would not be any transmission of the vibration and again, no wave motion. Although the above statements are meaningless in the case of waves that do not require a medium, they reveal a characteristic that is relevant to all waves regardless of origin: within a wave, the phase of a vibration (that is, its position within the vibration cycle) is different for adjacent points in space because the vibration reaches these points at different times.

Similarly, wave processes revealed from the study of waves other than sound waves can be significant to the understanding of sound phenomena. A relevant example is Thomas Young's principle of interference (Young, 1802, in Hunt 1992, p. 132). This principle was first introduced in Young's study of light and, within some specific contexts (for example, scattering of sound by sound), is still a researched area in the study of sound.

Mathematical Description of One-dimensional Waves

Wave Equation

Consider a travelling transverse wave (which may be a pulse) on a string (the medium). Consider the string to have a single spatial dimension. Consider this wave as travelling.

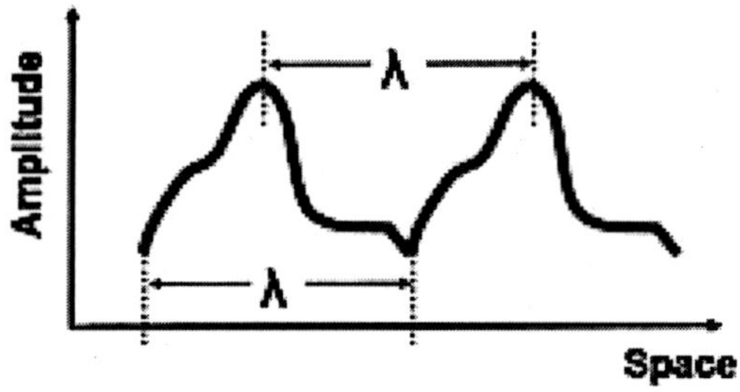

Figure: Wavelength λ, can be measured between any two corresponding points on a waveform

- In the x direction in space. E.g., let the positive x direction be to the right, and the negative x direction be to the left.
- With constant amplitude u
- With constant velocity v, where v is
- Independent of wavelength (no dispersion)
- Independent of amplitude (linear media, not nonlinear).
- With constant waveform, or shape.

This wave can then be described by the two-dimensional functions

$u(x, t) = F(x - v\,t)$ (waveform F travelling to the right)

$u(x, t) = G(x + v\,t)$ (waveform G travelling to the left)

or, more generally, by d'Alembert's formula:

$$u(x,t) = F(x - vt) + G(x + vt).$$

Representing two component waveforms F and G travelling through the medium in opposite directions. This wave can also be represented by the partial differential equation

$$\frac{1}{v^2}\frac{\partial^2 u}{\partial t^2} = \frac{\partial^2 u}{\partial x^2}.$$

General solutions are based upon Duhamel's principle.

Wave Forms

The form or shape of F in d'Alembert's formula involves the argument $x - vt$. Constant values of this argument correspond to constant values of F, and these constant values occur if x increases at the same rate that vt increases. That is, the wave shaped like the function F will move in the positive x-direction at velocity v (and G will propagate at the same speed in the negative x-direction). In the case of a periodic function F with period λ, that is, $F(x + \lambda - vt) = F(x - vt)$, the periodicity of F in space means that a snapshot of the wave at a given time t finds the wave varying periodically in space with period λ (the wavelength of the wave). In a similar fashion, this periodicity of F *implies* a periodicity in time as well: $F(x - v(t + T)) = F(x - vt)$ provided $vT = \lambda$, so an observation of the wave at a fixed location x finds the wave undulating periodically in time with period $T = \lambda/v$.

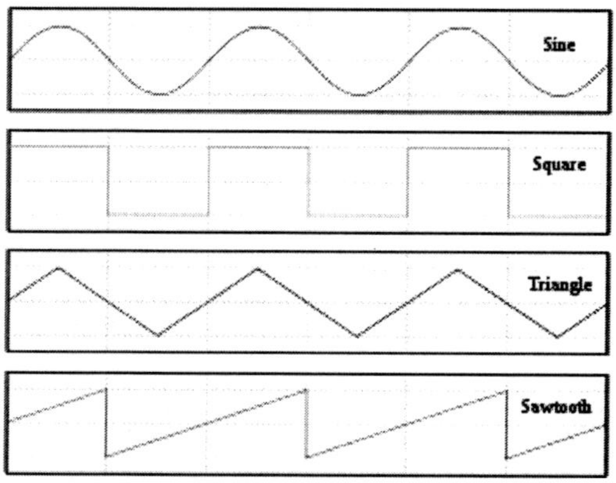

Figure: Sine, square, triangle and sawtooth waveforms.

Amplitude and Modulation

The amplitude of a wave may be constant (in which case the wave is a *c.w.* or *continuous wave*), or may be modulated so as to vary with time and/or position. The outline of the variation in amplitude is called the *envelope* of the wave. Mathematically, the modulated wave can be written in the form:

$$u(x,t) = A(x,t)\sin(kx - \omega t + \phi),$$

where $A(x,t)$ is the amplitude envelope of the wave, k is the *wavenumber* and φ is the *phase*. If the group velocity vg is wavelength-independent, this equation can be simplified as:

$$u(x,t) = A(x - v_g\, t)\sin(kx - \omega t + \phi),$$

showing that the envelope moves with the group velocity and retains its shape. Otherwise, in cases where the group velocity varies with wavelength, the pulse shape changes in a manner often described using an *envelope equation.*

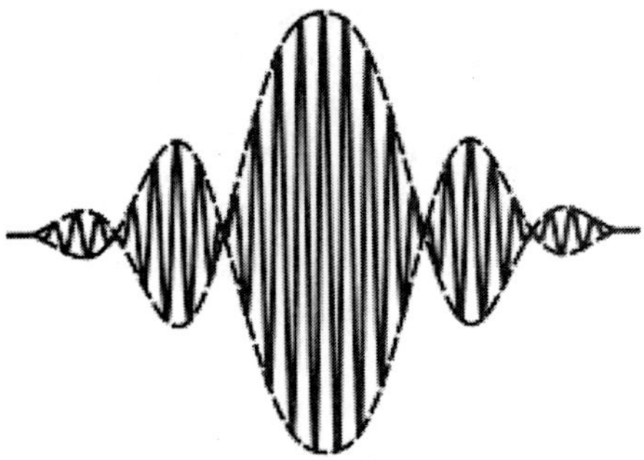

Figure: Illustration of the envelope (the slowly varying red curve) of an amplitude-modulated wave. The fast varying blue curve is the carrier wave, which is being modulated.

Phase Velocity and Group Velocity

There are two velocities that are associated with waves, the phase velocity and the group velocity. To understand them, one must consider several types of waveform. For simplification, examination is restricted to one dimension. The most basic wave (a form of plane wave) may be expressed in the form:

$$\psi(x,t) = Ae^{i(kx-\omega t)},$$

which can be related to the usual sine and cosine forms using Euler's formula. Rewriting the argument, $kx - \omega t = (\frac{2\pi}{\lambda})(x - vt)$, makes clear that this expression describes a vibration of wavelength $\lambda = \frac{2\pi}{k}$ travelling in the x-direction with a constant phase *velocity* $v_p = \frac{\omega}{k}$.

The other type of wave to be considered is one with localised structure described by an envelope, which may be expressed mathematically as, for example:

$$\psi(x,t) = \int_{-\infty}^{\infty} dk_1\ A(k_1)\ e^{i(k_1 x-\omega t)},$$

where now $A(k_1)$ (the integral is the inverse fourier transform of A(k1)) is a function exhibiting a sharp peak in a region of wave vectors Δk surrounding the point $k_1 = k$. In exponential form:

$$A = A_o(k_1)e^{i\alpha(k_1)},$$

with A_0 the magnitude of A. For example, a common choice for A_0 is a Gaussian wave packet:

$$A_o(k_1) = N\, e^{-\sigma^2 (k_1-k)^2/2} \, ,$$

where σ determines the spread of k_1-values about k, and N is the amplitude of the wave. The exponential function inside the integral for ψ oscillates rapidly with its argument, say $\varphi(k_1)$, and where it varies rapidly, the exponentials cancel each other out, interfere destructively, contributing little to ψ. However, an exception occurs at the location where the argument φ of the exponential varies slowly. (This observation is the basis for the method of stationary phase for evaluation of such integrals.) The condition for φ to vary slowly is that its rate of change with k_1 be small; this rate of variation is:

$$\Psi\psi \left.\frac{d\varphi}{dk_1}\right|_{k_1=k} = x - t\left.\frac{d\omega}{dk_1}\right|_{k_1=k} + \left.\frac{d\alpha}{dk_1}\right|_{k_1=k} ,$$

where the evaluation is made at $k_1 = k$ because $A(k_1)$ is centred there. This result shows that the position x where the phase changes slowly, the position where ψ is appreciable, moves with time at a speed called the *group velocity*:

$$v_g = \frac{d\omega}{dk} .$$

The group velocity therefore depends upon the dispersion relation connecting ω and k. For example, in quantum mechanics the energy of a particle represented as a wave packet is $E = \hbar\omega = (\hbar k)^2/(2m)$. Consequently, for that wave situation, the group velocity is

$$v_g = \frac{\hbar k}{m} ,$$

showing that the velocity of a localised particle in quantum mechanics is its group velocity. Because the group velocity varies with k, the shape of the wave packet broadens with time, and the particle becomes less localised. In other words, the velocity of the constituent waves of the wave packet travel at a rate that varies with their wavelength, so some move faster than others, and they cannot maintain the same interference pattern as the wave propagates.

Sinusoidal Waves

Mathematically, the most basic wave is the (spatially) one-dimensional sine wave (or *harmonic wave* or *sinusoid*) with an amplitude u described by the equation:

$$u(x, t) = A \sin(kx - \omega t + \phi) ,$$

where

- A is the maximum amplitude of the wave, maximum distance from the highest point of the disturbance in the medium (the crest) to the equilibrium point during one wave cycle. In the illustration to the right, this is the maximum vertical distance between the baseline and the wave.
- x is the space coordinate
- t is the time coordinate
- k is the wavenumber
- w is the angular frequency
- φ is the phase.

The units of the amplitude depend on the type of wave. Transverse mechanical waves (e.g., a wave on a string) have an amplitude expressed as a distance (e.g., metres), longitudinal mechanical waves (e.g., sound waves) use units of pressure (e.g., pascals), and electromagnetic waves (a form of transverse vacuum wave) express the amplitude in terms of its electric field (e.g., volts/metre).

The wavelength λ is the distance between two sequential crests or troughs (or other equivalent points), generally is measured in metres.

A wavenumber k, the spatial frequency of the wave in radians per unit distance (typically per metre), can be associated with the wavelength by the relation

$$k = \frac{2\pi}{\lambda}.$$

The period T is the time for one complete cycle of an oscillation of a wave. The frequency f is the number of periods per unit time (per second) and is typically measured in hertz. These are related by:

$$f = \frac{1}{T}.$$

In other words, the frequency and period of a wave are reciprocals.

The angular frequency ω represents the frequency in radians per second. It is related to the frequency or period by

$$\omega = 2\pi f = \frac{2\pi}{T}.$$

The wavelength λ of a sinusoidal waveform travelling at constant speed v is given by:

$$\lambda = \frac{v}{f},$$

where v is called the phase speed (magnitude of the phase velocity) of the wave and f is the wave's frequency. Wavelength can be a useful concept even if the wave is not periodic in space. For example, in an ocean wave approaching shore, the incoming wave undulates with a varying *local* wavelength that depends in part on the depth of the sea floor compared to the wave height. The analysis of the wave can be based upon comparison of the local wavelength with the local water depth.

Although arbitrary wave shapes will propagate unchanged in lossless linear time-invariant systems, in the presence of dispersion the sine wave is the unique shape that will propagate unchanged but for phase and amplitude, making it easy to analyse. Due to the Kramers–Kronig relations, a linear medium with dispersion also exhibits loss, so the sine wave propagating in a dispersive medium is attenuated in certain frequency ranges that depend upon the medium. The sine function is periodic, so the sine wave or sinusoid has a wavelength in space and a period in time.

The sinusoid is defined for all times and distances, whereas in physical situations we usually deal with waves that exist for a limited span in space and duration in time. Fortunately, an arbitrary wave shape can be decomposed into an infinite set of sinusoidal waves by the use of Fourier analysis. As a result, the simple case of a single sinusoidal wave can be applied to more general cases. In particular, many media are linear, or nearly so, so the calculation of arbitrary wave behaviour can be found by adding up responses to individual sinusoidal waves using the superposition principle to find the solution for a general waveform. When a medium is nonlinear, the response to complex waves cannot be determined from a sine-wave decomposition.

Standing Waves

A standing wave, also known as a *stationary wave*, is a wave that remains in a constant position. This phenomenon can occur because the medium is moving in the opposite direction to the wave, or it can arise in a stationary medium as a result of interference between two waves travelling in opposite directions.

The *sum* of two counter-propagating waves (of equal amplitude and frequency) creates a *standing wave*. Standing waves commonly arise when a boundary blocks further propagation of the wave, thus causing wave reflection, and therefore introducing a counter-propagating wave. For example when aviolin string is displaced, transverse waves propagate out to where the string is held in place at the bridge and the nut, where the waves are reflected back. At the bridge and nut, the two opposed waves are in antiphase and cancel each other, producing a node. Halfway between two nodes there is anantinode, where the two counter-propagating waves *enhance* each other maximally. There is no netpropagation of energy over time.

Physical Properties

Waves exhibit common behaviours under a number of standard situations.

Transmission and Media

Waves normally move in a straight line (i.e. rectilinearly) through a transmission *medium*. Such media can be classified into one or more of the following categories:

- A *bounded medium* if it is finite in extent, otherwise an unbounded *medium*
- A *linear medium* if the amplitudes of different waves at any particular point in the medium can be added
- A *uniform medium* or *homogeneous medium* if its physical properties are unchanged at different locations in space
- An *anisotropic medium* if one or more of its physical properties differ in one or more directions
- An *isotropic medium* if its physical properties are the *same* in all directions.

Reflection

When a wave strikes a reflective surface, it changes direction, such that the angle made by the incident wave and line normal to the surface equals the angle made by the reflected wave and the same normal line.

Interference

Waves that encounter each other combine through superposition to create a new wave called an interference pattern. Important interference patterns occur for waves that are in phase.

Reflection (Physics)

Reflection is the change in direction of a wavefront at an interface between two different media so that the wavefront returns into the medium from which it originated. Common examples include the reflection of light, sound and water waves. The *law of reflection* says that for specular reflection the angle at which the wave is incident on the surface equals the angle at which it is reflected. Mirrors exhibit specular reflection.

In acoustics, reflection causes echoes and is used in sonar. In geology, it is important in the study of seismic waves. Reflection is observed with surface waves in bodies of water. Reflection is observed with many types of electromagnetic wave, besides visible light. Reflection ofVHF and higher frequencies is important for radio transmission and for radar. Even hard X-rays and gamma rays can be reflected at shallow angles with special "grazing" mirrors.

Reflection of Light

Reflection of light is either *specular* (mirror-like) or *diffuse* (retaining the energy, but losing the image) depending on the nature of the interface. Furthermore, if the interface is between a dielectric and a conductor, the phase of the reflected wave is retained, otherwise if the interface is between two dielectrics, the phase may be retained or inverted, depending on the indices of refraction.

A mirror provides the most common model for specular light reflection, and typically consists of a glass sheet with a metallic coating where the reflection actually occurs. Reflection is enhanced in metals by suppression of wave propagation beyond their skin depths. Reflection also occurs at the surface of transparent media, such as water or glass.

In fact, reflection of light may occur whenever light travels from a medium of a given refractive index into a medium with a different refractive index. In the most general case, a certain fraction of the light is reflected from the interface, and the remainder is refracted. Solving Maxwell's equations for a light ray striking a boundary allows the derivation of the Fresnel equations, which can be used to predict how much of the light is reflected, and how much is refracted in a given situation. Total internal reflection of light from a denser medium occurs if the angle of incidence is above the critical angle.

Total internal reflection is used as a means of focusing waves that cannot effectively be reflected by common means. X-ray telescopes are constructed by creating a converging "tunnel" for the waves. As the waves interact at low angle with the surface of this tunnel they are reflected toward the focus point (or toward another interaction with the tunnel surface, eventually being directed to the detector at the focus). A conventional reflector would be useless as the X-rays would simply pass through the intended reflector.

When light reflects off a material denser (with higher refractive index) than the external medium, it undergoes a polarity inversion. In contrast, a less dense, lower refractive index material will reflect light in phase. This is an important principle in the field of thin-film optics.

Specular reflection forms images. Reflection from a flat surface forms a mirror image, which appears to be reversed from left to right because we compare the image we see to what we would see if we were rotated into the position of the image.

Specular reflection at a curved surface forms an image which may be magnified or demagnified; curved mirrors have optical power. Such mirrors may have surfaces that are spherical or parabolic.

Laws of Reflection

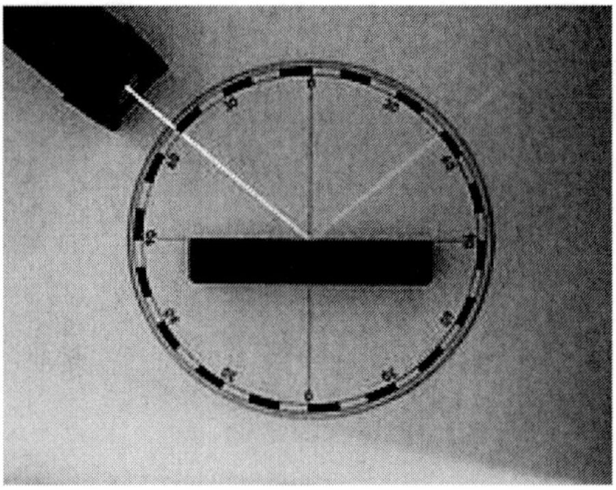

Figure: Specular reflection

If the reflecting surface is very smooth, the reflection of light that occurs is called specular or regular reflection. The laws of reflection are as follows:

1. The incident ray, the reflected ray and the normal to the reflection surface at the point of the incidence lie in the same plane.
2. The angle which the incident ray makes with the normal is equal to the angle which the reflected ray makes to the same normal.
3. The reflected ray and the incident ray are on the opposite sides of the normal.

Mechanism

In the classical electrodynamics, light is considered as electromagnetic wave, which is governed by the Maxwell Equations. Light waves incident on a material induce small oscillations of polarisation in the individual atoms (or oscillation of electrons, in metals), causing each particle to radiate a small secondary wave (in all directions, like a dipole antenna). All these waves add up to give specular reflection and refraction, according to the Huygens-Fresnel principle.

In case of dielectric (glass), the electric field of the light acts on the electrons in the glass, the moving electrons generate a field and become a new radiator.

The refraction light in the glass is the combined of the forward radiation of the electrons and the incident light and; the backward radiation is the one we see reflected from the surface of transparent materials, this radiation comes from everywhere in the glass, but it turns out that the total effect is equivalent to a reflection from the surface.

In metals, the electrons with no binding energy are called free electrons. The density number of the free electrons is very large. When these electrons oscillate with the incident light, the phase differences between the radiation field of these electrons and the incident field are ð, so the forward radiation will compensate the incident light at a skin depth, and backward radiation is just the reflected light.

Light–matter interaction in terms of photons is a topic of quantum electrodynamics, and is described in detail by Richard Feynman in his popular book *QED: The Strange Theory of Light and Matter*.

Diffuse Reflection

When light strikes the surface of a (non-metallic) material it bounces off in all directions due to multiple reflections by the microscopic irregularities *inside* the material (e.g. the grain boundaries of a polycrystalline material, or the cell or fibreboundaries of an organic material) and by its surface, if it is rough. Thus, an 'image' is not formed.

This is called *diffuse reflection*. The exact form of the reflection depends on the structure of the material.

One common model for diffuse reflection is Lambertian reflectance, in which the light is reflected with equal luminance (in photometry) or radiance (in radiometry) in all directions, as defined by Lambert's cosine law.

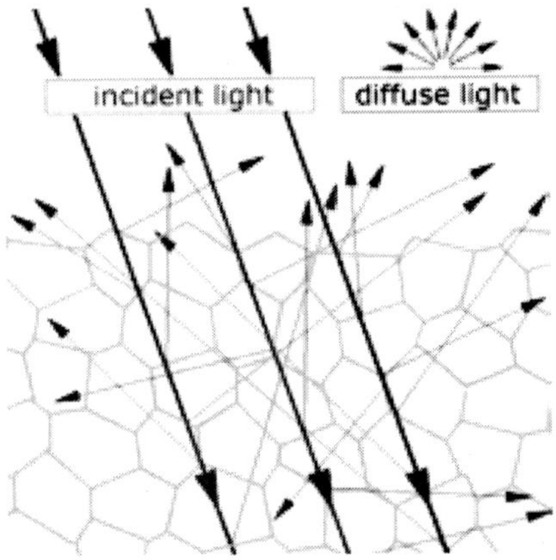

Figure: General scattering mechanism which gives diffuse reflection by a solid surface

The light sent to our eyes by most of the objects we see is due to diffuse reflection from their surface, so that this is our primary mechanism of physical observation.

Retroreflection

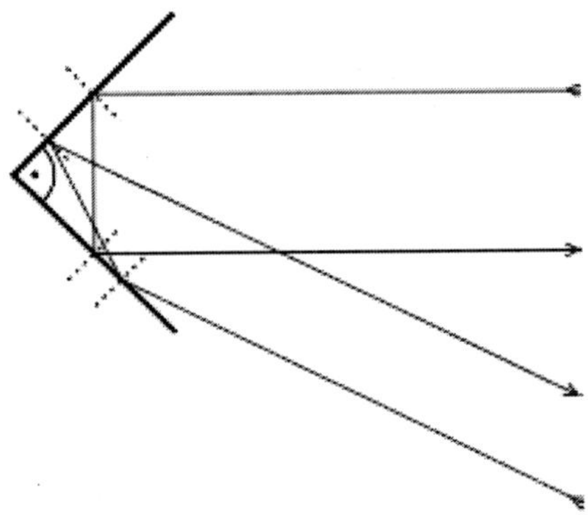

Figure: Working principle of a corner reflector

Some surfaces exhibit *retroreflection*. The structure of these surfaces is such that light is returned in the direction from which it came. When flying over clouds illuminated by sunlight the region seen around the aircraft's shadow will appear brighter, and a similar effect may be seen from dew on grass. This partial retro-reflection is created by the refractive properties of the curved droplet's surface and reflective properties at the backside of the droplet. Some animals' retinas act as retroreflectors, as this effectively improves the animals' night vision. Since the lenses of their eyes modify reciprocally the paths of the incoming and outgoing light the effect is that the eyes act as a strong retroreflector, sometimes seen at night when walking in wildlands with a flashlight. A simple retroreflector can be made by placing three ordinary mirrors mutually perpendicular to one another (a corner reflector). The image produced is the inverse of one produced by a single mirror. A surface can be made partially retroreflective by depositing a layer of tiny refractive spheres on it or by creating small pyramid like structures.

In both cases internal reflection causes the light to be reflected back to where it originated. This is used to make traffic signs and automobile license plates reflect light mostly back in the direction from which it came. In this application perfect retroreflection is not desired, since the light would then be directed back into the headlights of an oncoming car rather than to the driver's eyes.

Multiple Reflections

When light reflects off a mirror, one image appears. Two mirrors placed exactly face to face give the appearance of an infinite number of images along a straight line. The multiple images seen between two mirrors that sit at an angle to each other lie over a circle. The centre of that circle is located at the imaginary intersection of the mirrors. A square of four mirrors placed face to face give the appearance of an infinite number of images arranged in a plane. The multiple images seen between four mirrors assembling a pyramid, in which each pair of mirrors sits an angle to each other, lie over a sphere. If the base of the pyramid is rectangle shaped, the images spread over a section of a torus.

Complex Conjugate Reflection

Light bounces exactly back in the direction from which it came due to a nonlinear optical process. In this type of reflection, not only the direction of the light is reversed, but the actual wavefronts are reversed as well. A conjugate reflector can be used to remove aberrations from a beam by reflecting it and then passing the reflection through the aber rating optics a second time.

Other types of Reflection

Neutron Reflection

Materials that reflect neutrons, for example beryllium, are used in nuclear reactors and nuclear weapons. In the physical and biological sciences, the reflection of neutrons off of atoms within a material is commonly used to determine the material's internal structure.

Sound Reflection

Figure: Sound diffusion panel for high frequencies

When a longitudinal sound wave strikes a flat surface, sound is reflected in a coherent manner provided that the dimension of the reflective surface is large compared to the wavelength of the sound. Note that audible sound has a very wide frequency range (from 20 to about 17000 Hz), and thus a very wide range of wavelengths (from about 20 mm to 17 m). As a result, the overall nature of the reflection varies according to the texture and structure of the surface. For example, porous materials will absorb some energy, and rough materials (where rough is relative to the wavelength) tend to reflect in many directions—to scatter the energy, rather than to reflect it coherently. This leads into the field of architectural acoustics, because the nature of these reflections is critical to the auditory feel of a space. In the theory of exterior noise mitigation, reflective surface size mildly detracts from the concept of a noise barrier by reflecting some of the sound into the opposite direction.

Seismic Reflection

Seismic waves produced by earthquakes or other sources (such as explosions) may be reflected by layers within the Earth. Study of the deep reflections of waves generated by earthquakes has allowed seismologists to determine the layered structure of the Earth. Shallower reflections are used in reflection seismology to study the Earth's crust generally, and in particular to prospect for petroleum and natural gas deposits.

Refraction

Refraction is the change in direction of a wave due to a change in its speed. It is essentially a surface phenomenon . The phenomenon is mainly in governance to the law of conservation of energy. The proper explanation would be that due to change of medium, the phase velocity of the wave is changed but its frequency remains constant. This is most commonly observed when a wave passes from one medium to another at any angle other than 90° or 0°.

Refraction of light is the most commonly observed phenomenon, but any type of wave can refract when it interacts with a medium, for example when sound waves pass from one medium into another or when water waves move into water of a different depth. Refraction is described by Snell's law, which states that for a given pair of media and a wave with a single frequency, the ratio of the sines of the angle of incidence θ_1 and angle of refraction θ_2 is equivalent to the ratio of phase velocities (v_1 / v_2) in the two media, or equivalently, to the opposite ratio of the indices of refraction (n_2 / n_1):

$$\frac{\sin\theta_1}{\sin\theta_2} = \frac{v_1}{v_2} = \frac{n_2}{n_1}.$$

In general, the incident wave is partially refracted and partially reflected; the details of this behaviour are described by the Fresnel equations.

Explanation

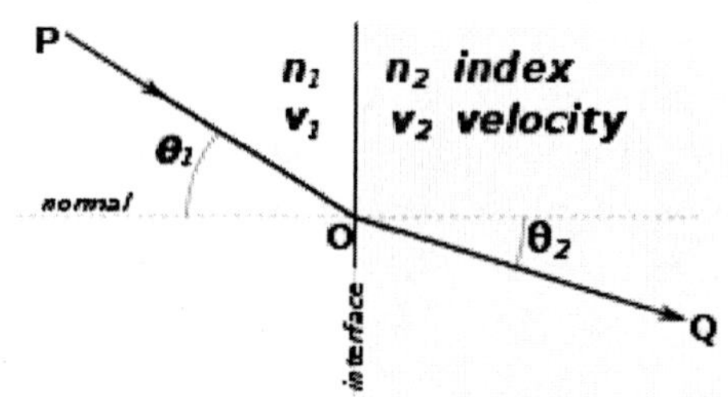

Figure: Refraction of light at the interface between two media of different refractive indices, with $n_2 > n_1$. Since the phase velocity is lower in the second medium ($v_2 < v_1$), the angle of refraction q_2 is less than the angle of incidence q_1; that is, the ray in the higher-index medium is closer to the normal.

In optics, refraction is a phenomenon that often occurs when waves travel from a medium with a given refractive index to a medium with another at an oblique angle. At the boundary between the media, the wave's phase velocity is altered, usually causing a change in direction. Its wavelength increases or decreases but its frequency remains constant. For example, a light ray will refract as it enters and leaves glass, assuming there is a change in refractive index. A ray travelling along the normal (perpendicular to the boundary) will change speed, but not direction. Refraction still occurs in this case. Understanding of this concept led to the invention of lenses and the refracting telescope.

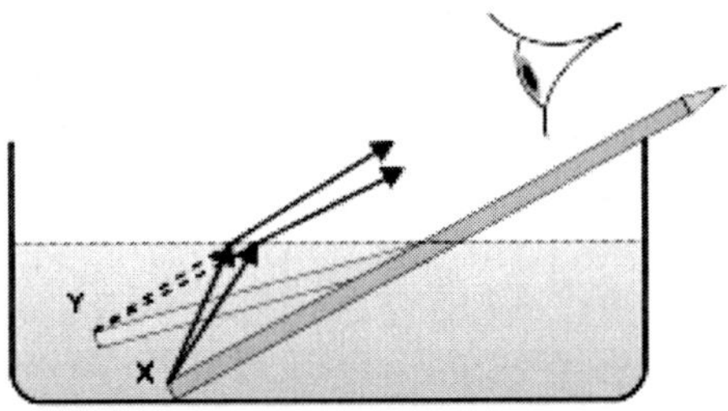

Figure: An object (in this case a pencil) part immersed in water looks bent due to refraction: the light waves from X change direction and so seem to originate at Y.

Refraction can be seen when looking into a bowl of water. Air has a refractive index of about 1.0003, and water has a refractive index of about 1.33. If a person looks at a straight object, such as a pencil or straw, which is placed at a slant, partially in the water, the object appears to bend at the water's surface. This is due to the bending of light rays as they move from the water to the air. Once the rays reach the eye, the eye traces them back as straight lines (lines of sight). The lines of sight (shown as dashed lines) intersect at a higher position than where the actual rays originated. This causes the pencil to appear higher and the water to appear shallower than it really is. The depth that the water appears to be when viewed from above is known as the *apparent depth*. This is an important consideration for spearfishing from the surface because it will make the target fish appear to be in a different place, and the fisher must aim lower to catch the fish.

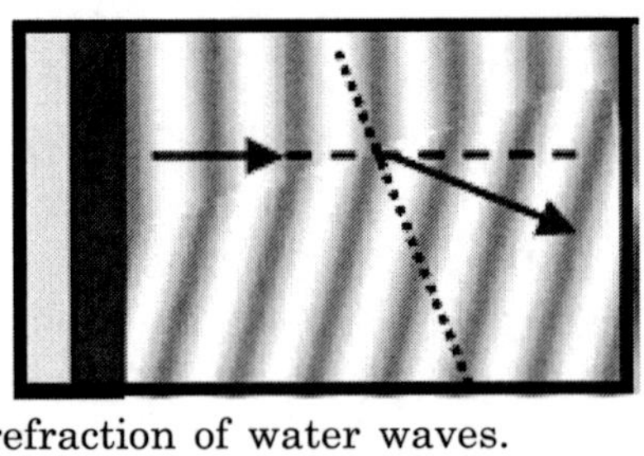

Figure: Diagram of refraction of water waves.

The diagram on the right shows an example of refraction in water waves. Ripples travel from the left and pass over a shallower region inclined at an angle to the wavefront. The waves travel slower in the more shallow water, so the wavelength decreases and the wave bends at the boundary. The dotted line represents the normal to the boundary. The dashed line represents the original direction of the waves. This phenomenon explains why waves on a shoreline tend to strike the shore close to a perpendicular angle. As the waves travel from deep water into shallower water near the shore, they are refracted from their original direction of travel to an angle more normal to the shoreline. Refraction is also responsible for rainbows and for the splitting of white light into a rainbow-spectrum as it passes through a glass prism. Glass has a higher refractive index than air. When a beam of white light passes from air into a material having an index of refraction that varies with frequency, a phenomenon known as dispersion occurs, in which different coloured components of the white light are refracted at different angles, i.e., they bend by different amounts at the interface, so that they become separated. The different colours correspond to different frequencies.

While refraction allows for beautiful phenomena such as rainbows, it may also produce peculiar optical phenomena, such as mirages and Fata Morgana. These are caused by the change of the refractive index of air with temperature.

Recently some metamaterials have been created which have a negative refractive index. With metamaterials, we can also obtain total refraction phenomena when the wave impedances of the two media are matched. There is then no reflected wave.

Also, since refraction can make objects appear closer than they are, it is responsible for allowing water to magnify objects. First, as light is entering a drop of water, it slows down. If the water's surface is not flat, then the light will be bent into a new path. This round shape will bend the light outwards and as it spreads out, the image you see gets larger.

A useful analogy in explaining the refraction of light would be to imagine a marching band as they march at an oblique angle from pavement (a fast medium) into mud (a slower medium). The marchers on the side that runs into the mud first will slow down first. This causes the whole band to pivot slightly toward the normal (make a smaller angle from the normal).

Clinical Significance

In medicine, particularly optometry, ophthalmology and orthoptics, *refraction* (also known as *refractometry*) is a clinical test in which a phoropter may be used by the appropriate eye care professional to determine the eye's refractive error and the best corrective lenses to be prescribed. A series of test lenses in graded optical powers or focal lengths are presented to determine which provide the sharpest, clearest vision.

Acoustics

In underwater acoustics, refraction is the bending or curving of a sound ray that results when the ray passes through a sound speed gradient from a region of one sound speed to a region of a different speed. The amount of ray bending is dependent upon the amount of difference between sound speeds, that is, the variation in temperature, salinity, and pressure of the water. Similar acoustics effects are also found in the Earth's atmosphere. The phenomenon of refraction of sound in the atmosphere has been known for centuries; however, beginning in the early 1970s, widespread analysis of this effect came into vogue through the designing of urban highways and noise barriers to address the meteorological effects of bending of sound rays in the lower atmosphere.

Diffraction

Diffraction refers to various phenomena which occur when a wave encounters an obstacle. Italian scientist Francesco Maria Grimaldi coined the word "diffraction" and was the first to record accurate observations of the phenomenon in 1665. In classical physics, the diffraction phenomenon is described as the apparent bending of waves around small obstacles and the spreading out of waves past small openings. Similar effects occur when light waves travel through a medium with a varying refractive index or a sound wave through one with varying acoustic impedance. Diffraction occurs with all waves, including sound waves, water waves, and electromagnetic waves such as visible light, x-rays and radio waves. As physical objects have wave-like properties (at the atomic level), diffraction also occurs with matter and can be studied according to the principles of quantum mechanics.

Richard Feynman said that "no-one has ever been able to define the difference between interference and diffraction satisfactorily. It is just a question of usage, and there is no specific, important physical difference between them."

He suggested that when there are only a few sources, say two, we call it interference, as in Young's slits, but with a large number of sources, the process is labelled diffraction.

While diffraction occurs whenever propagating waves encounter such changes, its effects are generally most pronounced for waves where the wavelength is roughly similar to the dimensions of the diffracting objects. If the obstructing object provides multiple, closely spaced openings, a complex pattern of varying intensity can result. This is due to the superposition, or interference, of different parts of a wave that travelled to the observer by different paths.

The formalism of diffraction can also describe the way in which waves of finite extent propagate in free space. For example, the expanding profile of a laser beam, the beam shape of a radar antenna and the field of view of an ultrasonic transducer can all be analysed using diffraction equations.

Examples

The effects of diffraction are often seen in everyday life. The most striking examples of diffraction are those involving light; for example, the closely spaced tracks on a CD or DVD act as a diffraction grating to form the familiar rainbow pattern seen when looking at a disk. This principle can be extended to engineer a grating with a structure such that it will produce any diffraction pattern desired; the hologram on a credit card is an example. Diffraction in the atmosphere by small particles can cause a bright ring to be visible around a bright light source like the sun or the moon. A shadow of a solid object, using light from a compact source, shows small fringes near its edges. The speckle pattern which is observed when laser light falls on an optically rough surface is also a diffraction phenomenon. All these effects are a consequence of the fact that light propagates as a wave.

Diffraction can occur with any kind of wave. Ocean waves diffract around jetties and other obstacles. Sound waves can diffract around objects, which is why one can still hear someone calling even when hiding behind a tree. Diffraction can also be a concern in some technical applications; it sets a fundamental limit to the resolution of a camera, telescope, or microscope.

History

The effects of diffraction of light were first carefully observed and characterised by Francesco Maria Grimaldi, who also coined the term *diffraction*, from the Latin *diffringere*, 'to break into pieces', referring to light breaking up into different directions. The results of Grimaldi's observations were published posthumously in 1665. Isaac Newton studied these effects and attributed them to *inflexion* of light rays.

James Gregory (1638–1675) observed the diffraction patterns caused by a bird feather, which was effectively the first diffraction grating to be discovered. Thomas Young performed a celebrated experiment in 1803 demonstrating interference from two closely spaced slits. Explaining his results by interference of the waves emanating from the two different slits, he deduced that light must propagate as waves. Augustin-Jean Fresnel did more definitive studies and calculations of diffraction, made public in 1815 and 1818, and thereby gave great support to the wave theory of light that had been advanced by Christiaan Huygens and reinvigorated by Young, against Newton's particle theory.

Mechanism

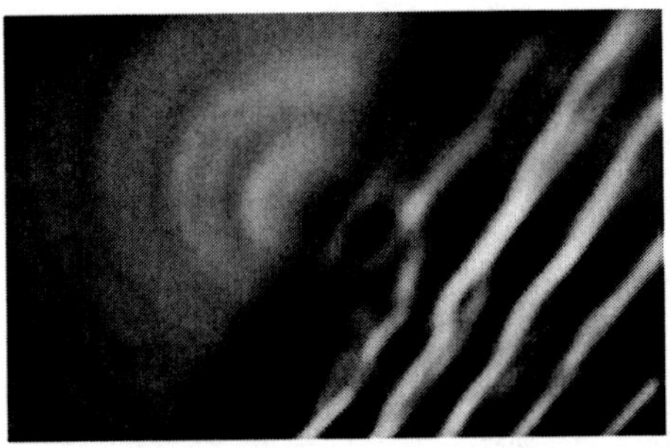

Figure: Photograph of single-slit diffraction in a circular ripple tank

Diffraction arises because of the way in which waves propagate; this is described by the Huygens–Fresnel principle and the principle of superposition of waves. The propagation of a wave can be visualised by considering every point on a wavefront as a point source for a secondary spherical wave. The wave displacement at any subsequent point is the sum of these secondary waves. When waves are added together, their sum is determined by the relative phases as well as the amplitudes of the individual waves so that the summed amplitude of the waves can have any value between zero and the sum of the individual amplitudes. Hence, diffraction patterns usually have a series of maxima and minima.

There are various analytical models which allow the diffracted field to be calculated, including the Kirchhoff-Fresnel diffraction equation which is derived from wave equation, the Fraunhofer diffraction approximation of the Kirchhoff equation which applies to the far field and the Fresnel diffraction approximation which applies to the near field. Most configurations cannot be solved analytically, but can yield numerical solutions through finite element and boundary element methods. It is possible to obtain a qualitative understanding of many diffraction phenomena by considering how the relative phases of the individual secondary wave sources vary, and in particular, the conditions in which the phase difference equals half a cycle in which case waves will cancel one another out.

The simplest descriptions of diffraction are those in which the situation can be reduced to a two-dimensional problem. For water waves, this is already the case; water waves propagate only on the surface of the water. For light, we can often neglect one direction if the diffracting object extends in that direction over a distance far greater than the wavelength. In the case of light shining through small circular holes we will have to take into account the full three dimensional nature of the problem.

Diffraction of Light

Some examples of diffraction of light are considered below.

Single-slit Diffraction

A long slit of infinitesimal width which is illuminated by light diffracts the light into a series of circular waves and the wavefront which emerges from the slit is a cylindrical wave of uniform intensity.

A slit which is wider than a wavelength produces interference effects in the space downstream of the slit. These can be explained by assuming that the slit behaves as though it has a large number of point sources spaced evenly across the width of the slit.

The analysis of this system is simplified if we consider light of a single wavelength. If the incident light is monochromatic, these sources all have the same phase. Light incident at a given point in the space downstream of the slit is made up of contributions from each of these point sources and if the relative phases of these contributions vary by 2π or more, we may expect to find minima and maxima in the diffracted light. Such phase differences are caused by differences in the path lengths over which contributing rays reach the point from the slit.

We can find the angle at which a first minimum is obtained in the diffracted light by the following reasoning. The light from a source located at the top edge of the slit interferes destructively with a source located at the middle of the slit, when the path difference between them is equal to $\lambda/2$. Similarly, the source just below the top of the slit will interfere destructively with the source located just below the middle of the slit at the same angle. We can continue this reasoning along the entire height of the slit to conclude that the condition for destructive interference for the entire slit is the same as the condition for destructive interference between two narrow slits a distance apart that is half the width of the slit. The path difference is given by $\frac{d\sin(\theta)}{2}$ so that the minimum intensity occurs at an angle θ_{min} given by

$$d\sin\theta_{\min} = \lambda$$

where

- d is the width of the slit,
- θ_{min} is the angle of incidence at which the minimum intensity occurs, and
- λ is the wavelength of the light

A similar argument can be used to show that if we imagine the slit to be divided into four, six, eight parts, etc., minima are obtained at angles θ_n given by

$$d\sin\theta_n = n\lambda$$

where

- n is an integer other than zero.

There is no such simple argument to enable us to find the maxima of the diffraction pattern. The intensity profile can be calculated using the Fraunhofer diffraction equation as

$$I(\theta) = I_0 \operatorname{sinc}^2(d\sin\theta / \lambda)$$

where

- $I(q)$ is the intensity at a given angle,
- I_0 is the original intensity, and
- The sinc function is given by $\operatorname{sinc}(x) = \sin(\Pi x)/(\Pi x)$ if $x \neq 0$, and $\operatorname{sinc}(0) = 1$

This analysis applies only to the far field, that is, at a distance much larger than the width of the slit.

Diffraction Grating

A diffraction grating is an optical component with a regular pattern. The form of the light diffracted by a grating depends on the structure of the elements and the number of elements present, but all gratings have intensity maxima at angles θ_m which are given by the grating equation

$$d\left(\sin\theta_m + \sin\theta_i\right) = m\lambda.$$

where

- θ_i is the angle at which the light is incident,
- *d* is the separation of grating elements, and
- *m* is an integer which can be positive or negative.

The light diffracted by a grating is found by summing the light diffracted from each of the elements, and is essentially a convolution of diffraction and interference patterns.

The figure shows the light diffracted by 2-element and 5-element gratings where the grating spacings are the same; it can be seen that the maxima are in the same position, but the detailed structures of the intensities are different.

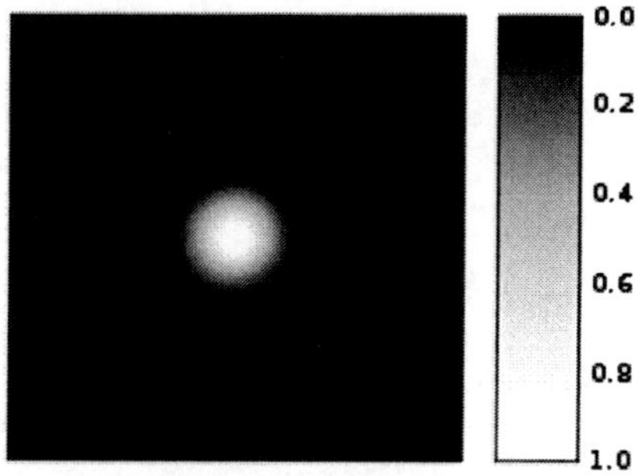

Figure: A computer-generated image of an Airy disk

Circular Aperture

The far-field diffraction of a plane wave incident on a circular aperture is often referred to as the Airy Disk. The variation in intensity with angle is given by

$$I(\theta) = I_0 \left(\frac{2J_1(ka\sin\theta)}{ka\sin\theta} \right)^2$$

where *a* is the radius of the circular aperture, *k* is equal to 2ð/λ and J_1 is a Bessel function. The smaller the aperture, the larger the spot size at a given distance, and the greater the divergence of the diffracted beams.

General Aperture

The wave that emerges from a point source has amplitudeψ at location r that is given by the solution of the frequency domain wave equation for a point source (The Helmholtz Equation),

$$\nabla^2\psi + k^2\psi = \delta(\mathbf{r})$$

where $\delta(\mathbf{r})$ is the 3-dimensional delta function. The delta function has only radial dependence, so the Laplace operator (aka scalar Laplacian) in the spherical coordinate system simplifies to.

$$\nabla^2\psi = \frac{1}{r}\frac{\partial^2}{\partial r^2}(r\psi)$$

By direct substitution, the solution to this equation can be readily shown to be the scalar Green's function, which in the spherical coordinate system (and using the physics time convention $e^{-i\omega t}$) is:

$$\psi(r) = \frac{e^{ikr}}{4\pi r}$$

This solution assumes that the delta function source is located at the origin. If the source is located at an arbitrary source point, denoted by the vector $\mathbf{r}'$ and the field point is located at the point $\mathbf{r}$, then we may represent the scalar Green's function (for arbitrary source location) as:

$$\psi(\mathbf{r}\,|\,\mathbf{r}') = \frac{e^{ik|\mathbf{r}-\mathbf{r}'|}}{4\pi\,|\mathbf{r}-\mathbf{r}'|}$$

Therefore, if an electric field, $E_{inc}(x,y)$ is incident on the aperture, the field produced by this aperture distribution is given by the surface integral:

$$\Psi(r) \propto \iint_{\text{aperture}} E_{inc}(x',y')\frac{e^{ik|\mathbf{r}-\mathbf{r}'|}}{4\pi\,|\mathbf{r}-\mathbf{r}'|}dx'dy',$$

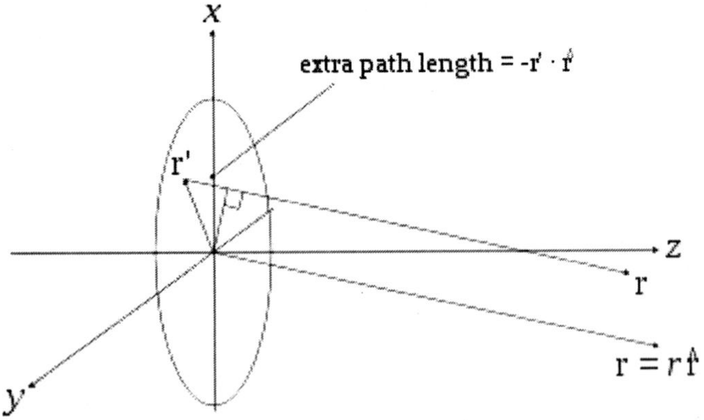

Figure: On the calculation of Fraunhofer region fields

where the source point in the aperture is given by the vector

$$\mathbf{r}' = x'\hat{\mathbf{x}} + y'\hat{\mathbf{y}}$$

In the far field, wherein the parallel rays approximation can be employed, the Green's function,

$$\psi(\mathbf{r}\,|\,\mathbf{r}') = \frac{e^{ik|\mathbf{r}-\mathbf{r}'|}}{4\pi\,|\mathbf{r}-\mathbf{r}'|}$$

simplifies to

$$\psi(\mathbf{r}\,|\,\mathbf{r}') = \frac{e^{ikr}}{4\pi r}e^{-ik(\mathbf{r}'\cdot\hat{\mathbf{r}})}$$

as can be seen in the figure to the right.

The expression for the far-zone (Fraunhofer region) field becomes

$$\Psi(r) \propto \frac{e^{ikr}}{4\pi r}\iint_{\text{aperture}} E_{inc}(x',y')e^{-ik(\mathbf{r}'\cdot\hat{\mathbf{r}})}\,dx'dy',$$

Now, since

$$\mathbf{r}' = x'\hat{\mathbf{x}} + y'\hat{\mathbf{y}}$$

and

$$\hat{\mathbf{r}} = \sin\theta\cos\phi\hat{\mathbf{x}} + \sin\theta\ \sin\phi\,\hat{\mathbf{y}} + \cos\theta\hat{\mathbf{z}}$$

the expression for the Fraunhofer region field from a planar aperture now becomes,

$$\Psi(r) \propto \frac{e^{ikr}}{4\pi r}\iint_{\text{aperture}} E_{inc}(x',y')e^{-ik\sin\theta(\cos\phi x'+\sin\phi y')}\,dx'dy'$$

Letting,

$$k_x = k\sin\theta\cos\phi$$

and

$$k_y = k\sin\theta\sin\phi$$

the Fraunhofer region field of the planar aperture assumes the form of a Fourier transform

$$\Psi(r) \propto \frac{e^{ikr}}{4\pi r}\iint_{\text{aperture}} E_{inc}(x',y')e^{-i(k_x x'+k_y y')}\,dx'dy',$$

In the far-field/Fraunhofer region, this becomes the spatial Fourier transform of the aperture distribution.

Huygens' principle when applied to an aperture simply says that the far-field diffraction pattern is the spatial Fourier transform of the aperture shape, and this is a direct by-product of using the parallel-rays approximation, which is identical to doing a plane wave decomposition of the aperture plane fields.

Propagation of a Laser Beam

The way in which the profile of a laser beam changes as it propagates is determined by diffraction. The output mirror of the laser is an aperture, and the subsequent beam shape is determined by that aperture. Hence, the smaller the output beam, the quicker it diverges.

Diode lasers have much greater divergence than He–Ne lasers for this reason.

Paradoxically, it is possible to reduce the divergence of a laser beam by first expanding it with one convex lens, and then collimating it with a second convex lens whose focal point is coincident with that of the first lens. The resulting beam has a larger aperture, and hence a lower divergence.

Diffraction-limited Imaging

The ability of an imaging system to resolve detail is ultimately limited by diffraction. This is because a plane wave incident on a circular lens or mirror is diffracted as described above. The light is not focused to a point but forms an Airy disk having a central spot in the focal plane with radius to first null of

$$d = 1.22\lambda N,$$

where l is the wavelength of the light and N is the f-number (focal length divided by diametre) of the imaging optics. In object space, the corresponding angular resolution is

$$\sin\theta = 1.22\frac{\lambda}{D},$$

where D is the diametre of the entrance pupil of the imaging lens (e.g., of a telescope's main mirror).

Two point sources will each produce an Airy pattern. As the point sources move closer together, the patterns will start to overlap, and ultimately they will merge to form a single pattern, in which case the two point sources cannot be resolved in the image.

The Rayleigh criterion specifies that two point sources can be considered to be resolvable if the separation of the two images is at least the radius of the Airy disk, i.e. if the first minimum of one coincides with the maximum of the other.

Thus, the larger the aperture of the lens, and the smaller the wavelength, the finer the resolution of an imaging system. This is why telescopes have very large lenses or mirrors, and why optical microscopes are limited in the detail which they can see.

Speckle Patterns

The speckle pattern which is seen when using a laser pointer is another diffraction phenomenon. It is a result of the superposition of many waves with different phases, which are produced when a laser beam illuminates a rough surface. They add together to give a resultant wave whose amplitude, and therefore intensity varies randomly.

Patterns

The upper half of this image shows a diffraction pattern of He-Ne laser beam on an elliptic aperture. The lower half is its 2D Fourier transform approximately reconstructing the shape of the aperture.

Several qualitative observations can be made of diffraction in general:

- The angular spacing of the features in the diffraction pattern is inversely proportional to the dimensions of the object causing the diffraction. In other words: The smaller the diffracting object, the 'wider' the resulting diffraction pattern, and vice versa. (More precisely, this is true of the sines of the angles.)
- The diffraction angles are invariant under scaling; that is, they depend only on the ratio of the wavelength to the size of the diffracting object.
- When the diffracting object has a periodic structure, for example in a diffraction grating, the features generally become sharper. The third figure, for example, shows a comparison of a double-slit pattern with a pattern formed by five slits, both sets of slits having the same spacing, between the centre of one slit and the next.

Particle Diffraction

Quantum theory tells us that every particle exhibits wave properties. In particular, massive particles can interfere and therefore diffract. Diffraction of electrons and neutrons stood as one of the powerful arguments in favour of quantum mechanics. The wavelength associated with a particle is the de Broglie wavelength

$$\lambda = \frac{h}{p}$$

where h is Planck's constant and p is the momentum of the particle (mass × velocity for slow-moving particles). For most macroscopic objects, this wavelength is so short that it is not meaningful to assign a wavelength to them. A sodium atom travelling at about 30,000 m/s would have a De Broglie wavelength of about 50 pico metres. Because the wavelength for even the smallest of macroscopic objects is extremely small, diffraction of matter waves is only visible for small particles, like electrons, neutrons, atoms and small molecules. The short wavelength of these matter waves makes them ideally suited to study the atomic crystal structure of solids and large molecules like proteins.

Relatively larger molecules like buckyballs were also shown to diffract.

Bragg Diffraction

Diffraction from a three dimensional periodic structure such as atoms in a crystal is called Bragg diffraction. It is similar to what occurs when waves are scattered from a diffraction grating. Bragg diffraction is a consequence of interference between waves reflecting from different crystal planes. The condition of constructive interference is given by *Bragg's law*:

$$m\lambda = 2d\sin\theta$$

where

λ is the wavelength,

d is the distance between crystal planes,

θ is the angle of the diffracted wave.

and m is an integer known as the *order* of the diffracted beam.

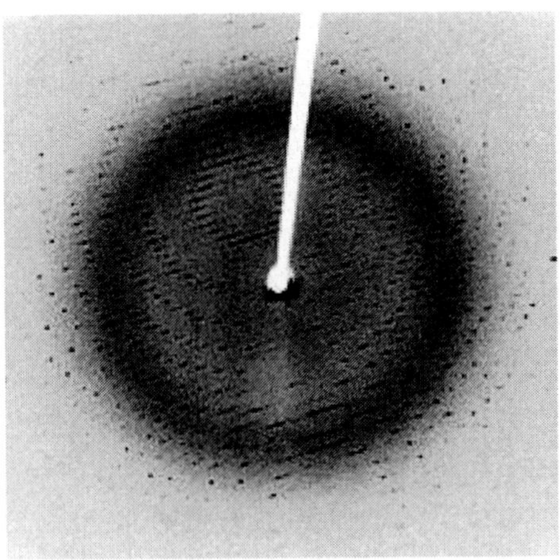

Figure: Following Bragg's law, each dot (or reflection), in this diffraction pattern forms from the constructive interference of X-rays passing through a crystal. The data can be used to determine the crystal's atomic structure.

Bragg diffraction may be carried out using either light of very short wavelength like x-rays or matter waves like neutrons (and electrons) whose wavelength is on the order of (or much smaller than) the atomic spacing.

The pattern produced gives information of the separations of crystallographic planes d, allowing one to deduce the crystal structure. Diffraction contrast, in electron microscopes and x-topography devices in particular, is also a powerful tool for examining individual defects and local strain fields in crystals.

Coherence

The description of diffraction relies on the interference of waves emanating from the same source taking different paths to the same point on a screen. In this description, the difference in phase between waves that took different paths is only dependent on the effective path length. This does not take into account the fact that waves that arrive at the screen at the same time were emitted by the source at different times. The initial phase with which the source emits waves can change over time in an unpredictable way.

This means that waves emitted by the source at times that are too far apart can no longer form a constant interference pattern since the relation between their phases is no longer time independent.

The length over which the phase in a beam of light is correlated, is called the coherence length. In order for interference to occur, the path length difference must be smaller than the coherence length. This is sometimes referred to as spectral coherence, as it is related to the presence of different frequency components in the wave. In the case of light emitted by an atomic transition, the coherence length is related to the lifetime of the excited state from which the atom made its transition.

If waves are emitted from an extended source, this can lead to incoherence in the transversal direction. When looking at a cross section of a beam of light, the length over which the phase is correlated is called the transverse coherence length. In the case of Young's double slit experiment, this would mean that if the transverse coherence length is smaller than the spacing between the two slits, the resulting pattern on a screen would look like two single slit diffraction patterns.

In the case of particles like electrons, neutrons and atoms, the coherence length is related to the spatial extent of the wave function that describes the particle.

Dispersion (Optics)

In optics, dispersion is the phenomenon in which the phase velocity of a wave depends on its frequency, or alternatively when the group velocity depends on the frequency. Media having such a property are termed *dispersive media*. Dispersion is sometimes called *chromatic dispersion* to emphasize its wavelength-dependent nature, or group-velocity dispersion(GVD) to emphasize the role of the group velocity.

The most familiar example of dispersion is probably a rainbow, in which dispersion causes the spatial separation of a white light into components of different wavelengths (different colours). However, dispersion also has an effect in many other circumstances: for example, GVD causes pulses to spread in optical fibres, degrading signals over long distances; also, a cancellation between group-velocity dispersion and nonlinear effects leads to soliton waves. Dispersion is most often described for light waves, but it may occur for any kind of wave that interacts with a medium or passes through an inhomogeneous geometry (e.g., a waveguide), such as sound waves.

There are generally two sources of dispersion: material dispersion and waveguide dispersion. Material dispersion comes from a frequency-dependent response of a material to waves. For example, material dispersion leads to undesired chromatic aberration in a lens or the separation of colours in a prism. Waveguide dispersion occurs when the speed of a wave in a waveguide (such as an optical fibre) depends on its frequency for geometric reasons, independent of any frequency dependence of the materials from which it is constructed. More generally, "waveguide" dispersion can occur for waves propagating through any inhomogeneous structure (e.g., a photonic crystal), whether or not the waves are confined to some region. In general, *both* types of dispersion may be present, although they are not strictly additive. Their combination leads to signal degradation in optical fibres for telecommunications, because the varying delay in arrival time between different components of a signal "smears out" the signal in time.

Material Dispersion in Optics

Material dispersion can be a desirable or undesirable effect in optical applications. The dispersion of light by glass prisms is used to construct spectrometres andspectroradiometres. Holographic gratings are also used, as they allow more accurate discrimination of wavelengths. However, in lenses, dispersion causes chromatic aberration, an undesired effect that may degrade images in microscopes, telescopes and photographic objectives.

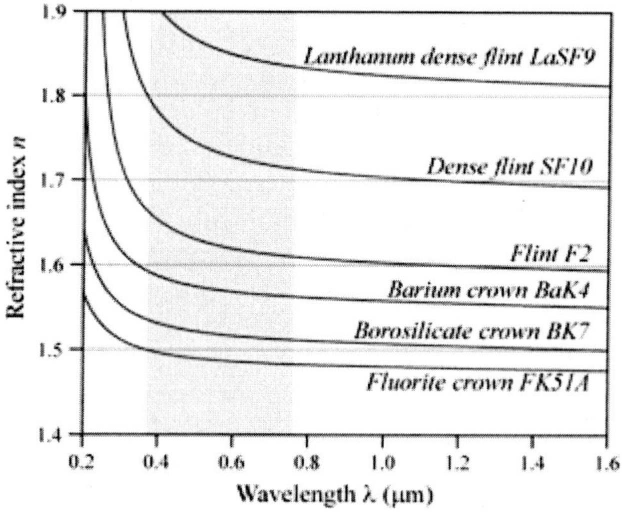

Figure: The variation of refractive index vs. vacuum wavelength for various glasses. The wavelengths of visible light are shaded in red.

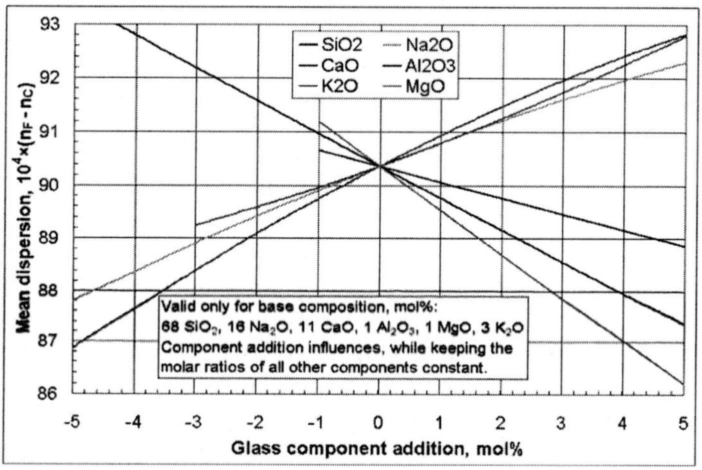

Figure: Influences of selected glass component additions on the mean dispersion of a specific base glass (n_F valid for λ = 486 nm (blue), n_C valid for λ = 656 nm (red))

The *phase velocity*, *v*, of a wave in a given uniform medium is given by

$$v = \frac{c}{n}$$

where *c* is the speed of light in a vacuum and *n* is the refractive index of the medium.

In general, the refractive index is some function of the frequency f of the light, thus $n = n(f)$, or alternatively, with respect to the wave's wavelength $n = n(\lambda)$. The wavelength dependence of a material's refractive index is usually quantified by its Abbe number or its coefficients in an empirical formula such as the Cauchy orSellmeier equations.

Because of the Kramers–Kronig relations, the wavelength dependence of the real part of the refractive index is related to the material absorption, described by the imaginary part of the refractive index (also called the extinction coefficient). In particular, for non-magnetic materials ($\mu = \mu_0$), the susceptibility ÷ that appears in the Kramers–Kronig relations is the electric susceptibility $\div_e = n^2 - 1$.

The most commonly seen consequence of dispersion in optics is the separation of white light into a colour spectrum by a prism. From Snell's law it can be seen that the angle of refraction of light in a prism depends on the refractive index of the prism material. Since that refractive index varies with wavelength, it follows that the angle that the light is refracted by will also vary with wavelength, causing an angular separation of the colours known as *angular dispersion.*

For visible light, refraction indices n of most transparent materials (e.g., air, glasses) decrease with increasing wavelength λ:

$$1 < n(\lambda_{\text{red}}) < n(\lambda_{\text{yellow}}) < n(\lambda_{\text{blue}}),$$

or alternatively:

$$\frac{dn}{d\lambda} < 0.$$

In this case, the medium is said to have *normal dispersion.* Whereas, if the index increases with increasing wavelength (which is typically the case for X-rays), the medium is said to have *anomalous dispersion.*

At the interface of such a material with air or vacuum (index of ~1), Snell's law predicts that light incident at an angle θ to the normal will be refracted at an angle $\arcsin(\sin(\theta)/n)$. Thus, blue light, with a higher refractive index, will be bent more strongly than red light, resulting in the well-known rainbow pattern.

Group and Phase Velocity

Another consequence of dispersion manifests itself as a temporal effect. The formula $v = c / n$ calculates the *phase velocity* of a wave; this is the velocity at which the *phase* of any one frequency component of the wave will propagate. This is not the same as the *group velocity* of the wave, that is the rate at which changes in amplitude (known as the *envelope* of the wave) will propagate. For a homogeneous medium, the group velocity v_g is related to the phase velocity by (here λ is the wavelength in vacuum, not in the medium):

$$v_g = c\left(n - \lambda \frac{dn}{d\lambda}\right)^{-1}.$$

The group velocity v_g is often thought of as the velocity at which energy or information is conveyed along the wave. In most cases this is true, and the group velocity can be thought of as the *signal velocity* of the waveform. In some unusual circumstances, called cases of anomalous dispersion, the rate of change of the index of refraction with respect to the wavelength changes sign, in which case it is possible for the group velocity to exceed the speed of light ($v_g > c$). Anomalous dispersion occurs, for instance, where the wavelength of the light is close to an absorption resonance of the medium. When the dispersion is anomalous, however, group velocity is no longer an indicator of signal velocity.

Instead, a signal travels at the speed of the wavefront, which is c irrespective of the index of refraction. Recently, it has become possible to create gases in which the group velocity is not only larger than the speed of light, but even negative. In these cases, a pulse can appear to exit a medium before it enters. Even in these cases, however, a signal travels at, or less than, the speed of light, as demonstrated by Stenner, et al.

The group velocity itself is usually a function of the wave's frequency. This results in group velocity dispersion (GVD), which causes a short pulse of light to spread in time as a result of different frequency components of the pulse travelling at different velocities. GVD is often quantified as the group *delay dispersion parameter* (again, this formula is for a uniform medium only):

$$D = -\frac{\lambda}{c}\frac{d^2 n}{d\lambda^2}.$$

If D is less than zero, the medium is said to have *positive dispersion*. If D is greater than zero, the medium has *negative dispersion*. If a light pulse is propagated through a normally dispersive medium, the result is the higher frequency components travel slower than the lower frequency components. The pulse therefore becomes *positively chirped*, or *up-chirped*, increasing in frequency with time. Conversely, if a pulse travels through an anomalously dispersive medium, high frequency components travel faster than the lower ones, and the pulse becomes *negatively chirped*, or *down-chirped*, decreasing in frequency with time.

The result of GVD, whether negative or positive, is ultimately temporal spreading of the pulse. This makes dispersion management extremely important in optical communications systems based on optical fibre, since if dispersion is too high, a group of pulses representing a bit-stream will spread in time and merge together, rendering the bit-stream unintelligible. This limits the length of fibre that a signal can be sent down without regeneration. One possible answer to this problem is to send signals down the optical fibre at a wavelength where the GVD is zero (e.g., around 1.3–1.5 μm in silica fibres), so pulses at this wavelength suffer minimal spreading from dispersion—in practice, however, this approach causes more problems than it solves because zero GVD unacceptably amplifies other nonlinear effects (such as four wave mixing). Another possible option is to use soliton pulses in the regime of anomalous dispersion, a form of optical pulse which uses a nonlinear optical effect to self-maintain its shape—solitons have the practical problem, however, that they require a certain power level to be maintained in the pulse for the nonlinear effect to be of the correct strength.

Instead, the solution that is currently used in practice is to perform dispersion compensation, typically by matching the fibre with another fibre of opposite-sign dispersion so that the dispersion effects cancel; such compensation is ultimately limited by nonlinear effects such as self-phase modulation, which interact with dispersion to make it very difficult to undo. Dispersion control is also important in lasers that produce short pulses. The overall dispersion of the optical resonator is a major factor in determining the duration of the pulses emitted by the laser. A pair of prisms can be arranged to produce net negative dispersion, which can be used to balance the usually positive dispersion of the laser medium. Diffraction gratings can also be used to produce dispersive effects; these are often used in high-power laser amplifier systems. Recently, an alternative to prisms and gratings has been developed: chirped mirrors. These dielectric mirrors are coated so that different wavelengths have different penetration lengths, and therefore different group delays. The coating layers can be tailored to achieve a net negative dispersion.

Dispersion in Waveguides

Optical fibres, which are used in telecommunications, are among the most abundant types of waveguides. Dispersion in these fibres is one of the limiting factors that determine how much data can be transported on a single fibre. The transverse modes for waves confined laterally within a waveguide generally have different speeds (and field patterns) depending upon their frequency (that is, on the relative size of the wave, the wavelength) compared to the size of the waveguide.

In general, for a waveguide mode with an angular frequency ω(β) at a propagation constant β (so that the electromagnetic fields in the propagation direction *(z)* oscillate proportional to $e^{i(\beta z - \omega t)}$), the group-velocity dispersion parameter D is defined as:

$$D = -\frac{2\pi c}{\lambda^2}\frac{d^2\beta}{d\omega^2} = \frac{2\pi c}{v_g^2\lambda^2}\frac{dv_g}{d\omega}$$

where $\lambda = 2\pi c / \omega$ is the vacuum wavelength and $v_g = d\omega / d\beta$ is the group velocity. This formula generalises the one in the previous section for homogeneous media, and includes both waveguide dispersion and material dispersion. The reason for defining the dispersion in this way is that $|D|$ is the (asymptotic) temporal pulse spreading Δt per unit bandwidth $\Delta\lambda$ per unit distance travelled, commonly reported in ps / nm km for optical fibres.

A similar effect due to a somewhat different phenomenon is modal dispersion, caused by a waveguide having multiple modes at a given frequency, each with a different speed. A special case of this is polarisation mode dispersion (PMD), which comes from a superposition of two modes that travel at different speeds due to random imperfections that break the symmetry of the waveguide.

Higher-order Dispersion Over Broad Bandwidths

When a broad range of frequencies (a broad bandwidth) is present in a single wavepacket, such as in an ultrashort pulse or a chirped pulse or other forms of spread spectrum transmission, it may not be accurate to approximate the dispersion by a constant over the entire bandwidth, and more complex calculations are required to compute effects such as pulse spreading.

In particular, the dispersion parameter D defined above is obtained from only one derivative of the group velocity. Higher derivatives are known as *higher-order dispersion.* These terms are simply a Taylor series expansion of the dispersion relation β(ω) of the medium or waveguide around some particular frequency. Their effects can be computed via numerical evaluation of Fourier transforms of the waveform, via integration of higher-order slowly varying envelope approximations, by a split-step method (which can use the exact dispersion relation rather than a Taylor series), or by direct simulation of the full Maxwell's equations rather than an approximate envelope equation.

Dispersion in Gemology

In the technical terminology of gemology, *dispersion* is the difference in the refractive index of a material at the B and G Fraunhofer wavelengths of 686.7 nm and 430.8 nm and is meant to express the degree to which a prism cut from the gemstone shows "fire", or colour. Dispersion is a material property. Fire depends on the dispersion, the cut angles, the lighting environment, the refractive index, and the viewer.

Dispersion in Imaging

In photographic and microscopic lenses, dispersion causes chromatic aberration, which causes the different colours in the image not to overlap properly. Various techniques have been developed to counteract this, such as the use of achromats, multielement lenses with glasses of different dispersion. They are constructed in such a way that the chromatic aberrations of the different parts cancel out.

Dispersion in Pulsar Timing

Pulsars are spinning neutron stars that emit pulses at very regular intervals ranging from milliseconds to seconds. Astronomers believe that the pulses are emitted simultaneously over a wide range of frequencies. However, as observed on Earth, the components of each pulse emitted at higher radio frequencies arrive before those emitted at lower frequencies. This dispersion occurs because of the ionised component of the interstellar medium, which makes the group velocity frequency dependent. The extra delay added at a frequency ν is

$$\pi t = k_{\mathrm{DM}} \times \left(\frac{\mathrm{DM}}{\nu^2} \right)$$

where the dispersion constant k_{DM} is given by

$$k_{\mathrm{DM}} = \frac{e^2}{2\pi m_e c} \,\square\, 4.149 \mathrm{GHz}^2 \mathrm{pc}^{-1} \mathrm{cm}^3 \mathrm{ms},$$

and the dispersion measure *DM* is the free electron column density (total electron content) *ne integrated* along the path travelled by the photon from the pulsar to the Earth, and is given by

$$\text{DM} = \int_0^d n_e \, dl$$

with units of parsecs per cubic centimetre (1pc/cm^3 = 30.857×10^{21} m^{-2}).

Typically for astronometric observations, this delay cannot be measured directly, since the emission time is unknown. What *can* be measured is the difference in arrival times at two different frequencies. The delay ΔT between a high frequency ν_{hi} and a low frequency ν_{lo} component of a pulse will be

$$\Delta t = k_{\text{DM}} \times \text{DM} \times \left(\frac{1}{\nu_{\text{lo}}^2} - \frac{1}{\nu_{\text{hi}}^2} \right)$$

Re-writing the above equation in terms of *DM* allows one to determine the *DM* by measuring pulse arrival times at multiple frequencies. This in turn can be used to study the interstellar medium, as well as allow for observations of pulsars at different frequencies to be combined.

Polarisation (Waves)

Polarisation (also polarisation) is a property of certain types of waves that describes the orientation of their oscillations. Electromagnetic waves, such as light, and gravitational waves exhibit polarisation; acoustic waves (sound waves) in a gas or liquid do not have polarisation because the direction of vibration and direction of propagation are the same.

By convention, the polarisation of light is described by specifying the orientation of the wave's electric field at a point in space over one period of the oscillation.

When light travels in free space, in most cases it propagates as a transverse wave—the polarisation is perpendicular to the wave's direction of travel.

In this case, the electric field may be oriented in a single direction (linear polarisation), or it may rotate as the wave travels (circular or elliptical polarisation).

In the latter cases, the oscillations can rotate either towards the right or towards the left in the direction of travel. Depending on which rotation is present in a given wave it is called the wave's chirality or handedness. In general the polarisation of an electromagnetic (EM) wave is a complex issue. For instance in a waveguide such as an optical fibre, or for radially polarised beams in free space, the description of the wave's polarisation is more complicated, as the fields can have longitudinal as well as transverse components. Such EM waves are either TM or hybrid modes.

For longitudinal waves such as sound waves in fluids, the direction of oscillation is by definition along the direction of travel, so there is no polarisation. In a solid medium, however, sound waves can be transverse. In this case, the polarisation is associated with the direction of the shear stress in the plane perpendicular to the propagation direction. This is important in seismology.

Polarisation is significant in areas of science and technology dealing with wave propagation, such as optics, seismology, telecommunications and radar science. The polarisation of light can be measured with a polarimetre. A polariser is a device that affects polarisation.

Theory

Basics: Plane Waves

The simplest manifestation of polarisation to visualise is that of a plane wave, which is a good approximation of most light waves (a plane wave is a wave with infinitely long and wide wavefronts).

For plane waves Maxwell's equations, specifically Gauss's laws, impose the transversality requirement that the electric and magnetic field be perpendicular to the direction of propagation and to each other. Conventionally, when considering polarisation, the electric field vector is described and the magnetic field is ignored since it is perpendicular to the electric field and proportional to it.

The electric field vector of a plane wave may be arbitrarily divided into two perpendicular components labeled x and y *(with* z indicating the direction of travel).

For a simple harmonic wave, where the amplitude of the electric vector varies in a sinusoidal manner in time, the two components have exactly the same frequency. However, these components have two other defining characteristics that can differ. First, the two components may not have the same amplitude. Second, the two components may not have the same phase, that is they may not reach their maxima and minima at the same time. Mathematically, the electric field of a plane wave can be written as,

$$v\vec{E}(\vec{r},t) = \mathrm{Re}\left[\left(A_x, A_y \cdot e^{i\phi}, 0\right)e^{i(kz-\omega t)}\right]$$

or alternatively,

$$\vec{E}(\vec{r},t) = (A_x \cos(kz - \omega t), A_y \cos(kz - \omega t + \phi), 0)$$

where A_x and A_y are the amplitudes of the x and y directions and φ is the relative phase between the two components.

Polarisation State

The shape traced out in a fixed plane by the electric vector as such a plane wave passes over it (a Lissajous figure) is a description of the polarisation state. The following figures show some examples of the evolution of the electric field vector (black), with time (the vertical axes), at a particular point in space, along with its *x* and *y* components (red/left and blue/right), and the path traced by the tip of the vector in the plane: The same evolution would occur when looking at the electric field at a particular time while evolving the point in space, along the direction opposite to propagation.

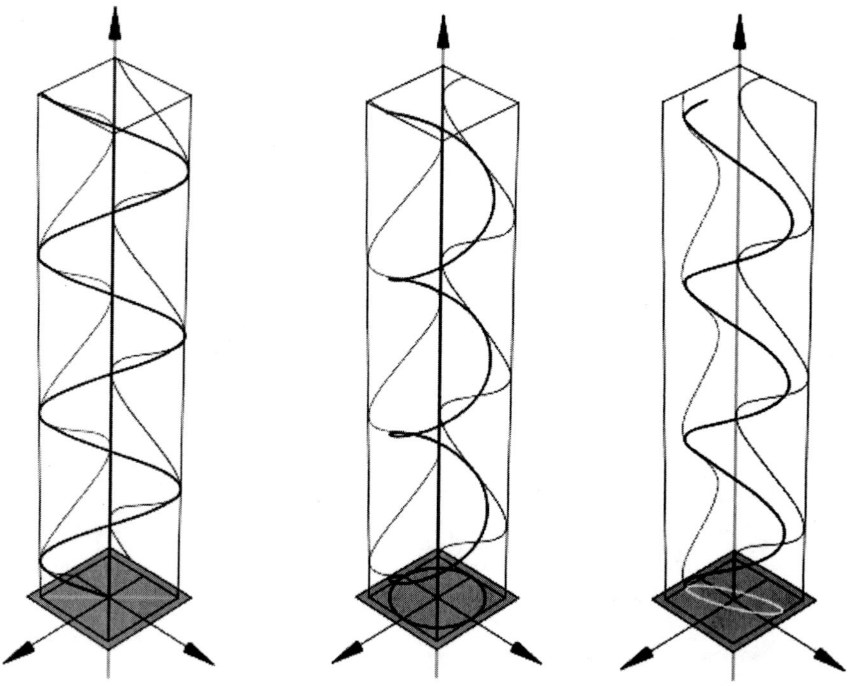

Linear Circular Elliptical

The two orthogonal (perpendicular) components are in phase. In this case the ratio of the strengths of the two components is constant, so the direction of the electric vector (the vector sum of these two components) is constant. Since the tip of the vector traces out a single line in the plane, this special case is called linear polarisation. The direction of this line depends on the relative amplitudes of the two components. The two orthogonal components have exactly the same amplitude and are exactly ninety degrees out of phase. In this case one component is zero when the other component is at maximum or minimum amplitude.

There are two possible phase relationships that satisfy this requirement: the x component can be ninety degrees ahead of the y component or it can be ninety degrees behind the y component. In this special case the electric vector traces out a circle in the plane, so this special case is called circular polarisation.

The direction the field rotates in depends on which of the two phase relationships exists. These cases are called *right-hand circular polarisation and left-hand circular polarisation*, depending on which way the electric vector rotates and the chosen convention.

Another case is when the two components are not in phase and either do not have the same amplitude or are not ninety degrees out of phase, though their phase offset and their amplitude ratio are constant. This kind of polarisation is called elliptical polarisation because the electric vector traces out an ellipse in the plane (the *polarisation ellipse*).

The "Cartesian" decomposition of the electric field into x and y components is, of course, arbitrary. Plane waves of any polarisation can be described instead by combining any two orthogonally polarised waves, for instance waves of opposite circular polarisation. The Cartesian polarisation decomposition is natural when dealing with reflection from surfaces, birefringent materials, or synchrotron radiation. The circularly polarised modes are a more useful basis for the study of light propagation in stereoisomers.

Though this section discusses polarisation for idealised plane waves, all the above is a very accurate description for most practical optical experiments which use TEM modes, including Gaussian optics.

Unpolarised Light

Most sources of electromagnetic radiation contain a large number of atoms or molecules that emit light. The orientation of the electric fields produced by these emitters may not be correlated, in which case the light is said to be *unpolarised.* If there is partial correlation between the emitters, the light is *partially polarised.* If the polarisation is consistent across the spectrum of the source, partially polarised light can be described as a superposition of a completely unpolarised component, and a completely polarised one. One may then describe the light in terms of the degree of polarisation, and the parameters of the polarisation ellipse.

Parameterisation

For ease of visualisation, polarisation states are often specified in terms of the polarisation ellipse, specifically its orientation and elongation. A common parameterisation uses the orientation angle, ψ, the angle between the major semi-axis of the ellipse and the x-axis (also known as tilt angle or azimuth angle) and the ellipticity, å, the major-to-minor-axis ratio (also known as the axial ratio). An ellipticity of zero or infinity corresponds to linear polarisation and an ellipticity of 1 corresponds to circular polarisation. The ellipticity angle, ÷ = arccot ε= arctan $1/\varepsilon$, is also commonly used. An example is shown in the diagram to the right. An alternative to the ellipticity or ellipticity angle is the eccentricity, however unlike the azimuth angle and ellipticity angle, the latter has no obvious geometrical interpretation in terms of the Poincaré sphere.

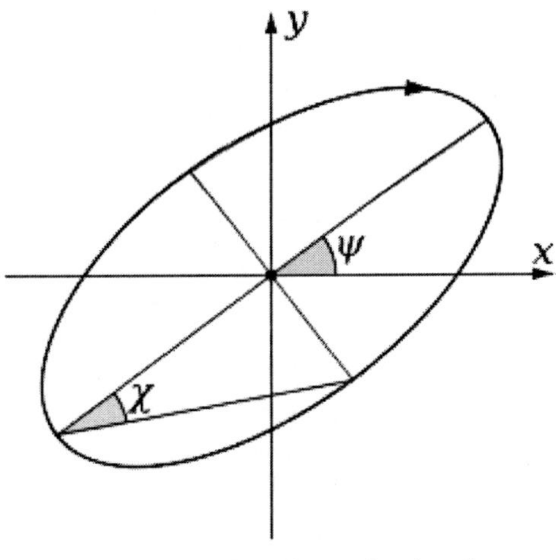

Full information on a completely polarised state is also provided by the amplitude and phase of oscillations in two components of the electric field vector in the plane of polarisation. This representation was used above to show how different states of polarisation are possible. The amplitude and phase information can be conveniently represented as a two-dimensional complex vector (the Jones vector):

$$\mathbf{e} = \begin{bmatrix} a_1 e^{i\theta_1} \\ a_2 e^{i\theta_2} \end{bmatrix}.$$

Here a_1 and a_2 denote the amplitude of the wave in the two components of the electric field vector, while θ_1 and θ_2 represent the phases. The product of a Jones vector with a complex number of unitmodulus gives a different Jones vector representing the same ellipse, and thus the same state of polarisation. The physical electric field, as the real part of the Jones vector, would be altered but the polarisation state itself is independent of absolute phase. The basis vectors used to represent the Jones vector need not represent linear polarisation states (i.e. be real). In general any two orthogonal states can be used, where an orthogonal vector pair is formally defined as one having a zero inner product. A common choice is left and right circular polarisations, for example to model the different propagation of waves in two such components in circularly birefringent media or signal paths of coherent detectors sensitive to circular polarisation. Regardless of whether polarisation ellipses are represented using geometric parameters or Jones vectors, implicit in the parameterisation is the orientation of the coordinate frame. This permits a degree of freedom, namely rotation about the propagation direction. When considering light that is propagating parallel to the surface of the Earth, the terms "horizontal" and "vertical" polarisation are often used, with the former being associated with the first component of the Jones vector, or zero azimuth angle. On the other hand, in astronomy the equatorial coordinate system is generally used instead, with the zero azimuth (or position angle, as it is more commonly called in astronomy to avoid confusion with the horizontal coordinate system) corresponding to due north.

S and P Polarisation

Another coordinate system frequently used relates to the plane made by the propagation direction and a vector perpendicular to the plane of a reflecting surface. This is known as the *plane of incidence.* The component of the electric field parallel to this plane is termed *p-like* (parallel) and the component perpendicular to this plane is termed *s-like* (from *senkrecht,* German for perpendicular). Light with a p-like electric field is said to be *p-polarised, pi-polarised, tangential plane polarised,* or is said to be a*transverse-magnetic* (TM) wave. Light with an s-like electric field is *s-polarised,* also known as *sigma-polarised* or *sagittal plane polarised,* or it can be called a *transverse-electric* (TE) wave. However, there is no universal convention in this TE and TM naming scheme, and certain authors do refer to light with p-like electric field as TE and light with s-like electric field as TM.

Parameterisation of Incoherent or Partially Polarised Radiation

In the case of partially-polarised radiation, the Jones vector varies in time and space in a way that differs from the constant rate of phase rotation of monochromatic, purely-polarised waves. In this case, the wave field is likely stochastic, and only statistical information can be gathered about the variations and correlations between components of the electric field. This information is embodied in the coherency matrix:

$$\mathbf{\Psi} = \left\langle \mathbf{e}\mathbf{e}^{\dagger} \right\rangle$$

$$= \left\langle \begin{bmatrix} e_1 e_1^* & e_1 e_2^* \\ e_2 e_1^* & e_2 e_2^* \end{bmatrix} \right\rangle$$

$$= \left\langle \begin{bmatrix} a_1^2 & a_1 a_2 e^{i(\theta_1 - \theta_2)} \\ a_1 a_2 e^{-i(\theta_1 - \theta_2)} & a_2^2 \end{bmatrix} \right\rangle$$

where angular brackets denote averaging over many wave cycles. Several variants of the coherency matrix have been proposed: the Wiener coherency matrix and the spectral coherency matrix of Richard Barakat measure the coherence of a spectral decomposition of the signal, while the Wolfcoherency matrix averages over all time/frequencies.

The coherency matrix contains all second order statistical information about the polarisation. This matrix can be decomposed into the sum of two idempotent matrices, corresponding to the eigenvectors of the coherency matrix, each representing a polarisation state that is orthogonal to the other. An alternative decomposition is into completely polarised (zero determinant) and unpolarised (scaled identity matrix) components. In either case, the operation of summing the components corresponds to the incoherent superposition of waves from the two components. The latter case gives rise to the concept of the "degree of polarisation"; i.e., the fraction of the total intensity contributed by the completely polarised component.

The coherency matrix is not easy to visualise, and it is therefore common to describe incoherent or partially polarised radiation in terms of its total intensity (I), (fractional) degree of polarisation (p), and the shape parameters of the polarisation ellipse. An alternative and mathematically convenient description is given by the Stokes parameters, introduced by George Gabriel Stokes in 1852. The relationship of the Stokes parameters to intensity and polarisation ellipse parameters is shown in the equations.

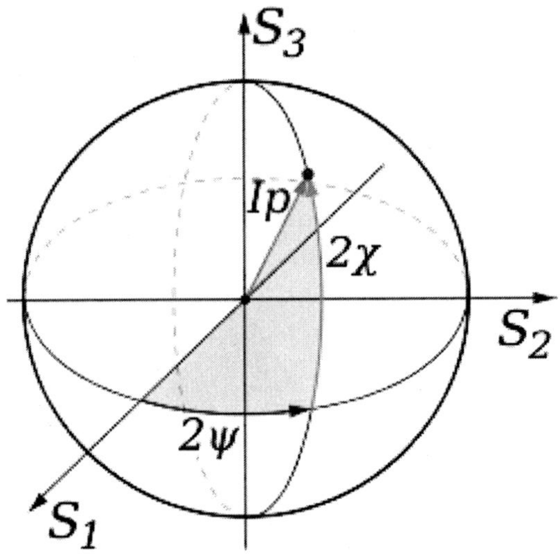

Figure: Poincaré sphere diagram

$$S_0 = I$$
$$S_1 = Ip\cos 2\psi \cos 2\chi$$
$$S_2 = Ip\sin 2\psi \cos 2\chi$$
$$S_3 = Ip\sin 2\chi$$

Here *Ip*, 2ψ and 2÷ are the spherical coordinates of the polarisation state in the three-dimensional space of the last three Stokes parameters. Note the factors of two before ψ and ÷ corresponding respectively to the facts that any polarisation ellipse is indistinguishable from one rotated by 180°, or one with the semi-axis lengths swapped accompanied by a 90° rotation. The Stokes parameters are sometimes denoted *I*, *Q*, *U* and *V*. The Stokes parameters contain all of the information of the coherency matrix, and are related to it linearly by means of the identity matrix plus the three Pauli matrices:

$$\mathbf{\Psi} = \frac{1}{2}\sum_{j=0}^{3} S_j \boldsymbol{\sigma}_j, \text{ where}$$

$$\boldsymbol{\sigma}_0 = \begin{bmatrix} 1 & 0 \\ 0 & 1 \end{bmatrix} \quad \boldsymbol{\sigma}_1 = \begin{bmatrix} 1 & 0 \\ 0 & -1 \end{bmatrix}$$

$$\boldsymbol{\sigma}_2 = \begin{bmatrix} 0 & 1 \\ 1 & 0 \end{bmatrix} \quad \boldsymbol{\sigma}_3 = \begin{bmatrix} 0 & -i \\ i & 0 \end{bmatrix}$$

Mathematically, the factor of two relating physical angles to their counterparts in Stokes space derives from the use of second-order moments and correlations, and incorporates the loss of information due to absolute phase invariance.

The figure above makes use of a convenient representation of the last three Stokes parameters as components in a three-dimensional vector space. This space is closely related to the Poincaré sphere, which is the spherical surface occupied by completely polarised states in the space of the vector

$$\mathbf{u} = \frac{1}{S_0}\begin{bmatrix} S_1 \\ S_2 \\ S_3 \end{bmatrix}.$$

All four Stokes parameters can also be combined into the four-dimensional Stokes vector, which can be interpreted as four-vectors of Minkowski space. In this case, all physically realisable polarisation states correspond to time-like, future-directed vectors.

Propagation, Reflection and Scattering

In a vacuum, the components of the electric field propagate at the speed of light, so that the phase of the wave varies in space and time while the polarisation state does not. That is,

$$\mathbf{e}(z + \Delta z, t + \Delta t) = \mathbf{e}(z, t)e^{ik(c\Delta t - \Delta z)},$$

where k is the wavenumber and positive z is the direction of propagation. As noted above, the physical electric vector is the real part of the Jones vector. When electromagnetic waves interact with matter, their propagation is altered. If this depends on the polarisation states of the waves, then their polarisation may also be altered.

In many types of media, electromagnetic waves may be decomposed into two orthogonal components that encounter different propagation effects. A similar situation occurs in the signal processing paths of detection systems that record the electric field directly. Such effects are most easily characterised in the form of a complex 2×2 transformation matrix called the Jones matrix:

$$\mathbf{e}' = \mathbf{J}\mathbf{e}.$$

In general the Jones matrix of a medium depends on the frequency of the waves.

For propagation effects in two orthogonal modes, the Jones matrix can be written as

$$\mathbf{J} = \mathbf{T}\begin{bmatrix} g_1 & 0 \\ 0 & g_2 \end{bmatrix}\mathbf{T}^{-1},$$

where g_1 and g_2 are complex numbers representing the change in amplitude and phase caused in each of the two propagation modes, and T is a unitary matrix representing a change of basis from these propagation modes to the linear system used for the Jones vectors. For those media in which the amplitudes are unchanged but a differential phase delay occurs, the Jones matrix is unitary, while those affecting amplitude without phase have Hermitian Jones matrices. In fact, since *any* matrix may be written as the product of unitary and positive Hermitian matrices, any sequence of linear propagation effects, no matter how complex, can be written as the product of these two basic types of transformations.

Media in which the two modes accrue a differential delay are called *birefringent*. Well known manifestations of this effect appear in optical wave plates/retarders (linear modes) and in Faraday rotation/optical rotation (circular modes). An easily visualised example is one where the propagation modes are linear, and the incoming radiation is linearly polarised at a 45° angle to the modes. As the phase difference starts to appear, the polarisation becomes elliptical, eventually changing to purely circular polarisation (90° phase difference), then to elliptical and eventually linear polarisation (180° phase) with an azimuth angle perpendicular to the original direction, then through circular again (270° phase), then elliptical with the original azimuth angle, and finally back to the original linearly polarised state (360° phase) where the cycle begins anew.

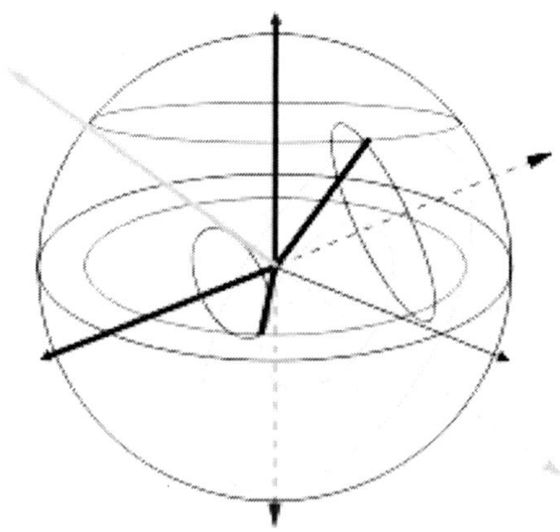

Figure: Paths taken by vectors in the Poincaré sphere under birefringence. The propagation modes (rotation axes) are shown with red, blue, and yellow lines, the initial vectors by thick black lines, and the paths they take by coloured ellipses (which represent circles in three dimensions).

In general the situation is more complicated and can be characterised as a rotation in the Poincaré sphere about the axis defined by the propagation modes (this is a consequence of the isomorphism of SU(2) with SO(3)). Examples for linear (blue), circular (red), and elliptical (yellow) birefringence are shown. The total intensity and degree of polarisation are unaffected. If the path length in the birefringent medium is sufficient, plane waves will exit the material with a significantly different propagation direction, due to refraction. For example, this is the case with macroscopic crystals of calcite, which present the viewer with two offset, orthogonally polarised images of whatever is viewed through them.

It was this effect that provided the first discovery of polarisation, by Erasmus Bartholinusin 1669. In addition, the phase shift, and thus the change in polarisation state, is usually frequency dependent, which, in combination with dichroism, often gives rise to bright colours and rainbow-like effects. Media in which the amplitude of waves propagating in one of the modes is reduced are called *dichroic*. Devices that block nearly all of the radiation in one mode are known as *polarising filters* or simply "polarisers". In terms of the Stokes parameters, the total intensity is reduced while vectors in the Poincaré sphere are "dragged" towards the direction of the favoured mode. Mathematically, under the treatment of the Stokes parameters as a Minkowski 4-vector, the transformation is a scaled Lorentz boost (due to the isomorphism of SL(2,C) and the restricted Lorentz group, SO(3,1)). Just as the Lorentz transformation preserves the proper time, the quantity det $\psi = S_0^2 - S_1^2 - S_2^2 - S_3^2$ is invariant within a multiplicative scalar constant under Jones matrix transformations (dichroic and/or birefringent).

In birefringent and dichroic media, in addition to writing a Jones matrix for the net effect of passing through a particular path in a given medium, the evolution of the polarisation state along that path can be characterised as the (matrix) product of an infinite series of infinitesimal steps, each operating on the state produced by all earlier matrices. In a uniform medium each step is the same, and one may write

$$\mathbf{J} = Je^{\mathbf{D}},$$

where J is an overall (real) gain/loss factor. Here D is a traceless matrix such that αDe gives the derivative of e with respect to z. If D is Hermitian the effect is dichroism, while a unitary matrix models birefringence. The matrix D can be expressed as a linear combination of the Pauli matrices, where real coefficients give Hermitian matrices and imaginary coefficients give unitary matrices. The Jones matrix in each case may therefore be written with the convenient construction

$$\mathbf{J_b} = J_b e^{\beta\sigma\cdot\hat{\mathbf{n}}} \quad \text{and} \quad \mathbf{J_r} = J_r e^{\phi i\sigma\cdot\hat{\mathbf{m}}},$$

where σ is a 3-vector composed of the Pauli matrices (used here as generators for the Lie groupSL(2,C)) and n and m are real 3-vectors on the Poincaré sphere corresponding to one of the propagation modes of the medium. The effects in that space correspond to a Lorentz boost of velocity parameter 2β along the given direction, or a rotation of angle 2φ about the given axis. These transformations may also be written as biquaternions (quaternions with complex elements), where the elements are related to the Jones matrix in the same way that the Stokes parameters are related to the coherency matrix. They may then be applied in pre- and post-multiplication to the quaternion representation of the coherency matrix, with the usual exploitation of the quaternion exponential for performing rotations and boosts taking a form equivalent to the matrix exponential equations above.

In addition to birefringence and dichroism in extended media, polarisation effects describable using Jones matrices can also occur at (reflective) interface between two materials of different refractive index. These effects are treated by the Fresnel equations. Part of the wave is transmitted and part is reflected, with the ratio depending on angle of incidence and the angle of refraction. In addition, if the plane of the reflecting surface is not aligned with the plane of propagation of the wave, the polarisation of the two parts is altered. In general, the Jones matrices of the reflection and transmission are real and diagonal, making the effect similar to that of a simple linear polariser. For unpolarised light striking a surface at a certain optimum angle of incidence known as Brewster's angle, the reflected wave will be completely *s*-polarised.

Certain effects do not produce linear transformations of the Jones vector, and thus cannot be described with (constant) Jones matrices. For these cases it is usual instead to use a 4×4 matrix that acts upon the Stokes 4-vector. Such matrices were first used by Paul Soleillet in 1929, although they have come to be known as Mueller matrices. While every Jones matrix has a Mueller matrix, the reverse is not true. Mueller matrices are frequently used to study the effects of the scattering of waves from complex surfaces or ensembles of particles.

Polarisation in Nature, Science, and Technology

Polarisation Effects in Everyday Life

Light reflected by shiny transparent materials is partly or fully polarised, except when the light is perpendicular to the surface. It was through this effect that polarisation was first discovered in 1808 by the mathematician Etienne Louis Malus. A polarising filter, such as a pair of polarising sunglasses, can be used to observe this effect by rotating the filter while looking through it at the reflection off of a distant horizontal surface. At certain rotation angles, the reflected light will be reduced or eliminated. Polarising filters remove light polarised at 90° to the filter's polarisation axis. If two polarisers are placed atop one another at 90° angles to one another, there is minimal light transmission.

Polarisation by scattering is observed as light passes through the atmosphere. The scattered light produces the brightness and colour in clear skies. This partial polarisation of scattered light can be used to darken the sky in photographs, increasing the contrast. This effect is easiest to observe at sunset, on the horizon at a 90° angle from the setting sun. Another easily observed effect is the drastic reduction in brightness of images of the sky and clouds reflected from horizontal surfaces, which is the main reason polarising filters are often used in sunglasses. Also frequently visible through polarising sunglasses are rainbow-like patterns caused by colour-dependent birefringent effects, for example in toughened glass (e.g., car windows) or items made from transparent plastics. The role played by polarisation in the operation of liquid crystal displays (LCDs) is also frequently apparent to the wearer of polarising sunglasses, which may reduce the contrast or even make the display unreadable.

The photograph on the right was taken through polarising sunglasses and through the rear window of a car. Light from the sky is reflected by the windshield of the other car at an angle, making it mostly horizontally polarised. The rear window is made of tempered glass. Stress from heat treatment of the glass alters the polarisation of light passing through it, like a wave plate. Without this effect, the sunglasses would block the horizontally polarised light reflected from the other car's window. The stress in the rear window, however, changes some of the horizontally polarised light into vertically polarised light that can pass through the glasses. As a result, the regular pattern of the heat treatment becomes visible.

Biology

Many animals are apparently capable of perceiving some of the components of the polarisation of light, e.g. linear horizontally-polarised light. This is generally used for navigational purposes, since the linear polarisation of sky light is always perpendicular to the direction of the sun. This ability is very common among the insects, including bees, which use this information to orient their communicative dances. Polarisation sensitivity has also been observed in species of octopus, squid, cuttlefish, and mantis shrimp. In the latter case, one species measures all six orthogonal components of polarisation, and is believed to have optimal polarisation vision. The rapidly changing, vividly coloured skin patterns of cuttlefish, used for communication, also incorporate polarisation patterns, and mantis shrimp are known to have polarisation selective reflective tissue. Sky polarisation was thought to be perceived by pigeons, which was assumed to be one of their aids in homing, but research indicates this is a popular myth. The naked human eye is weakly sensitive to polarisation, without the need for intervening filters. Polarised light creates a very faint pattern near the centre of the visual field, called Haidinger's brush. This pattern is very difficult to see, but with practice one can learn to detect polarised light with the naked eye.

Geology

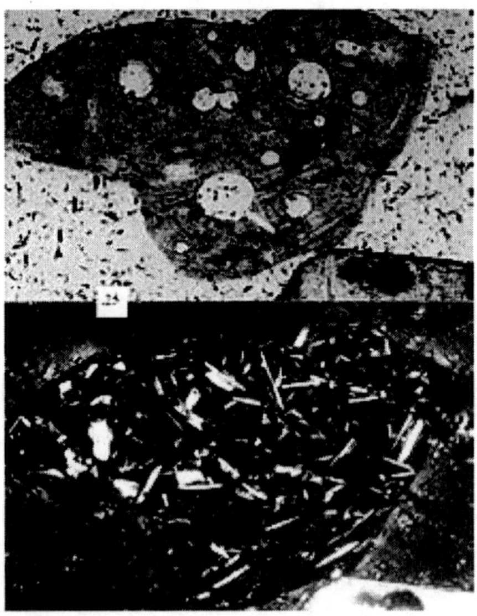

Figure: Photomicrograph of a volcanic sand grain; upper picture is plane-polarised light, bottom picture is cross-polarised light, scale box at left-centre is 0.25 millimetre.

The property of (linear) birefringence is widespread in crystalline minerals, and indeed was pivotal in the initial discovery of polarisation. In mineralogy, this property is frequently exploited using polarisation microscopes, for the purpose of identifying minerals.

Chemistry

Polarisation is principally of importance in chemistry due to circular dichroism and "optical rotation" (circular birefringence) exhibited by optically active (chiral) molecules. It may be measured using polarimetry.

The term "polarisation" may also refer to the through-bond (inductive or resonant effect) or through-space influence of a nearby functional group on the electronic properties (e.g.dipole moment) of a covalent bond or atom. This concept is based on the formation of an electric dipole within a molecule, which is related to polarisation of electromagnetic waves in infrared spectroscopy. Molecules will absorb infrared light if the frequency of the bond vibration is resonant with (identical to) the incident light frequency, where the molecular vibration at hand produces a change in the dipole moment of the molecule. In some nonlinear optical processes, the direction of an oscillating dipole will dictate the polarisation of the emitted electromagnetic radiation, as in vibrational sum frequency generation spectroscopy or similar processes.

Polarised light does interact with anisotropic materials, which is the basis for birefringence. This is usually seen in crystalline materials and is especially useful in geology. The polarised light is "double refracted", as the refractive index is different for horizontally and vertically polarised light in these materials. This is to say, the polarisability of anisotropic materials is not equivalent in all directions. This anisotropy causes changes in the polarisation of the incident beam, and is easily observable using cross-polar microscopy or polarimetry. The optical rotation of chiral compounds (as opposed to achiral compounds that form anisotropic crystals), is derived from circular birefringence. Like linear birefringence described above, circular birefringence is the "double refraction" of circular polarised light.

Astronomy

In many areas of astronomy, the study of polarised electromagnetic radiation from outer space is of great importance. Although not usually a factor in the thermal radiation of stars, polarisation is also present in radiation from coherent astronomical sources (e.g. hydroxyl or methanol masers), and incoherent sources such as the large radio lobes in active galaxies, and pulsar radio radiation (which may, it is speculated, sometimes be coherent), and is also imposed upon starlight by scattering from interstellar dust. Apart from providing information on sources of radiation and scattering, polarisation also probes the interstellar magnetic field via Faraday rotation. The polarisation of the cosmic microwave background is being used to study the physics of the very early universe. Synchrotron radiation is inherently polarised. It has been suggested that astronomical sources caused the chirality of biological molecules on Earth.

3D movies

Polarisation is also used for some 3D movies, in which the images intended for each eye are either projected from two different projectors with orthogonally oriented polarising filters or, more typically, from a single projector with time multiplexed polarisation (a fast alternating polarisation device for successive frames). Polarised 3D glasses with suitable polarised filters ensure that each eye receives only the intended image.

Historical stereoscopic projection displays used linear polarisation encoding because it was inexpensive and offered good separation. Circular polarisation makes left-eye/right-eye separation insensitive to the viewing orientation; circular polarisation is used in typical 3-D movie exhibition today, such as the system from RealD. Polarised 3-D only works on screens that maintain polarisation (such as silver screens); a normal projection screen would cause depolarisation which would void the effect.

Communication and Radar Applications

All radio transmitting and receiving antennas are intrinsically polarised, special use of which is made in radar. Most antennas radiate either horizontal, vertical, or circular polarisation although elliptical polarisation also exists. The electric field or E-plane determines the polarisation or orientation of the radio wave. Vertical polarisation is most often used when it is desired to radiate a radio signal in all directions such as widely distributed mobile units. AM and FM radio use vertical polarisation, while television uses horizontal polarisation. Alternating vertical and horizontal polarisation is used on satellite communications (including television satellites), to allow the satellite to carry two separate transmissions on a given frequency, thus doubling the number of customers a single satellite can serve. Electronically controlled birefringent devices are used in combination with polarising filters as modulators in fibre optics.

Materials Science

In engineering, the relationship between strain and birefringence motivates the use of polarisation in characterising the distribution of stress and strain in prototypes.

Navigation

Sky polarisation has been exploited in the "sky compass", which was used in the 1950s when navigating near the poles of the Earth's magnetic field when neither the sun norstars were visible (e.g. under daytime cloud or twilight). It has been suggested, controversially, that the Vikings exploited a similar device (the "sunstone") in their extensive expeditions across the North Atlantic in the 9th–11th centuries, before the arrival of the magnetic compass in Europe in the 12th century. Related to the sky compass is the "polar clock", invented by Charles Wheatstone in the late 19th century.

Explanation of Cognitive Illusions

Perceptual Organisation

To make sense of the world it is necessary to organise incoming sensations into information which is meaningful. Gestalt psychologists believe one way this is done is by perceiving individual sensory stimuli as a meaningful whole. Gestalt organisation can be used to explain many illusions including the Duck-Rabbit illusion where the image as a whole switches back and forth from being a duck then being a rabbit and why in the figure-ground illusion the figure and ground are reversible.

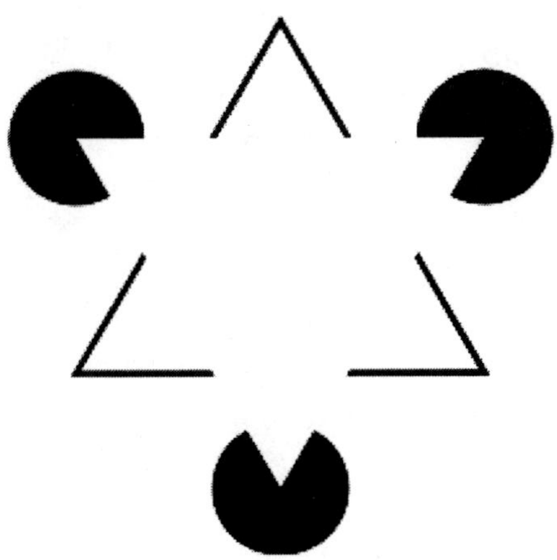

Figure: Kanizsa triangle

In addition, Gestalt theory can be used to explain the illusory contours in the Kanizsa Triangle. A floating white triangle, which does not exist, is seen. The brain has a need to see familiar simple objects and has a tendency to create a "whole" image from individual elements. Gestalt means "form" or "shape" in German. However, another explanation of the Kanizsa Triangle is based in evolutionary psychology and the fact that in order to survive it was important to see form and edges. The use of perceptual organisation to create meaning out of stimuli is the principle behind other well-known illusions including impossible objects. Our brain makes sense of shapes and symbols putting them together like a jigsaw puzzle, formulating that which isn't there to that which is believable.

Depth and Motion Perception

Illusions can be based on an individual's ability to see in three dimensions even though the image hitting the retina is only two dimensional. The Ponzo illusion is an example of an illusion which uses monocular cues of depth perception to fool the eye.

In the Ponzo illusion the converging parallel lines tell the brain that the image higher in the visual field is farther away therefore the brain perceives the image to be larger, although the two images hitting the retina are the same size. The Optical illusion seen in a diorama/false perspective also exploits assumptions based on monocular cues of depth perception. The M. C. Escher painting*Waterfall* exploits rules of depth and proximity and our understanding of the physical world to create an illusion. Like depth perception, motion perception is responsible for a number of sensory illusions. Film animation is based on the illusion that the brain perceives a series of slightly varied images produced in rapid succession as a moving picture. Likewise, when we are moving, as we would be while riding in a vehicle, stable surrounding objects may appear to move. We may also perceive a large object, like an aeroplane, to move more slowly than smaller objects, like a car, although the larger object is actually moving faster. The Phi phenomenon is yet another example of how the brain perceives motion, which is most often created by blinking lights in close succession.

Colour and Brightness Constancies

Perceptual constancies are sources of illusions. Colour constancy and brightness constancy are responsible for the fact that a familiar object will appear the same colour regardless of the amount of or colour of light reflecting from it.

Figure: Simultaneous Contrast Illusion. The background is a colour gradient and progresses from dark grey to light grey. The horizontal bar appears to progress from light grey to dark grey, but is in fact just one colour.

An illusion of colour or contrast difference can be created when the luminosity or colour of the area surrounding an unfamiliar object is changed.

The contrast of the object will appear darker against a black field that reflects less light compared to a white field even though the object itself did not change in colour. Similarly, the eye will compensate for colour contrast depending on the colour cast of the surrounding area.

Figure: In this illusion, the coloured regions appear rather different, roughly orange and brown. In fact they are the same colour, and in identical immediate surrounds, but the brain changes its assumption about colour due to the global interpretation of the surrounding image. Also, the white tiles that are shadowed are the same colour as the grey tiles outside the shadow.

Object Consistencies

Just as it perceives colour and brightness constancies, the brain has the ability to understand familiar objects as having a consistent shape or size. For example a door is perceived as rectangle regardless of how the image may change on the retina as the door is opened and closed. Unfamiliar objects, however, do not always follow the rules of shape constancy and may change when the perspective is changed. The Shepard illusion of the changing table is an example of an illusion based on distortions in shape constancy.

Future Perception

Researcher Mark Changizi of Rensselaer Polytechnic Institute in New York has a more imaginative take on optical illusions, saying that they are due to a neural lag which most humans experience while awake. When light hits the retina, about one-tenth of a second goes by before the brain translates the signal into a visual perception of the world. Scientists have known of the lag, yet they have debated how humans compensate, with some proposing that our motor system somehow modifies our movements to offset the delay. Changizi asserts that the human visual system has evolved to compensate for neural delays by generating images of what will occur one-tenth of a second into the future. This foresight enables humans to react to events in the present, enabling humans to perform reflexive acts like catching a fly ball and to maneuver smoothly through a crowd. Illusions occur when our brains attempt to perceive the future, and those perceptions don't match reality. For example, an illusion called the Hering illusionlooks like bicycle spokes around a central point, with vertical lines on either side of this central, so-called vanishing point. The illusion tricks us into thinking we are moving forward, and thus, switches on our future-seeing abilities. Since we aren't actually moving and the figure is static, we misperceive the straight lines as curved ones.

Changizi Said:

"Evolution has seen to it that geometric drawings like this elicit in us premonitions of the near future. The converging lines toward a vanishing point (the spokes) are cues that trick our brains into thinking we are moving forward—as we would in the real world, where the door frame (a pair of vertical lines) seems to bow out as we move through it—and we try to perceive what that world will look like in the next instant."

Cognitive Processes Hypothesis

The hypothesis claims that visual illusions occur because the neural circuitry in our visual system evolves, by neural learning, to a system that makes very efficient interpretations of usual 3D scenes based in the emergence of simplified models in our brain that speed up the interpretation process but give rise to optical illusions in unusual situations. In this sense, the cognitive processes hypothesis can be considered a framework for an understanding of optical illusions as the signature of the empirical statistical way vision has evolved to solve the inverse problem.

Research indicates that 3D vision capabilities emerge and are learned jointly with the planning of movements. After a long process of learning, an internal representation of the world emerges that is well adjusted to the perceived data coming from closer objects. The representation of distant objects near the horizon is less "adequate". In fact, it is not only the Moon that seems larger when we perceive it near the horizon. In a photo of a distant scene, all distant objects are perceived as smaller than when we observe them directly using our vision.

The retinal image is the main source driving vision but what we see is a "virtual" 3D representation of the scene in front of us. We don't see a physical image of the world; we see objects, and the physical world is not itself separated into objects. We see it according to the way our brain organises it. The names, colours, usual shapes and other information about the things we see pop up instantaneously from our neural circuitry and influence the representation of the scene. We "see" the most relevant information about the elements of the best 3D image that our neural networks can produce. The illusions arise when the "judgments" implied in the unconscious analysis of the scene are in conflict with reasoned considerations about it.

Afterimage

An afterimage or ghost image or image burn-in is an optical illusion that refers to an image continuing to appear in one's vision after the exposure to the original image has ceased. One of the most common afterimages is the bright glow that seems to float before one's eyes after looking into a light source for a few seconds.

Closing the eye can help achieve a better sense of the colour in its own aspect.

Afterimages come in two forms, negative (inverted) and positive (retaining original colour). The process behind positive afterimages is unknown, though thought to be related to neural adaptation. On the other hand, negative afterimages are a retinal phenomenon and are well understood.

Negative Afterimages

Negative afterimages are caused when the eye's photoreceptors, primarily those known as cone cells, adapt from the overstimulation and lose sensitivity. Normally the eye deals with this problem by rapidly moving small amounts, the motion later being "filtered out" so it is not noticeable.

However if the colour image is large enough that the small movements are not enough to change the colour under one area of the retina, those cones will eventually tire or adapt and stop responding. The rod cells can also be affected by this.

When the eyes are then diverted to a blank space, the adapted photoreceptors send out a weak signal and those colours remain muted. However, the surrounding cones that were not being excited by that colour are still "fresh", and send out a strong signal. The signal is exactly the same as if looking at the opposite colour, which is how the brain interprets it. Ewald Hering explained how the brain sees afterimages, in terms of three pairs of primary colours.

Thisopponent process theory states that the human visual system interprets colour information by processing signals from cones and rods in an antagonistic manner. The opponent colour theory suggests that there are three opponent channels: red versus cyan, blue versus yellow, and black versus white. Responses to one colour of an opponent channel are antagonistic to those to the other colour. Therefore, a green image will produce a magenta afterimage. The green colour tires out the green photoreceptors, so they produce a weaker signal. Anything resulting in less green, is interpreted as its paired primary colour, which is magenta.

Afterimage on Empty Shape

An afterimage in general is an optical illusion that refers to an image continuing to appear after exposure to the original image has ceased. Prolonged viewing of the coloured patch induces an afterimage of the complementary colour (for example, yellow colour induces a bluish afterimage). "After image on empty shape" effect is related also to a class of effects related to Contrast effect. In this effect, an empty (white) shape is presented on a coloured background for several seconds. When the background colour disappears (becomes white) an illusionary colour, similar to the original background is perceived within the shape. The mechanism of the effect is still unclear, and may be produced by one or two of the following mechanisms:

- During the presentation of the empty shape on a coloured background, the coloured background induces an illusionary complementary colour "Induced Colour" inside the empty shape (i.e. bird). After the disappearance of the coloured background an afterimage of the "Induced Colour" might appear inside the "empty shape". Thus, the expected colour of the shape will be complementary to the "Induced Colour", and therefore similar to the colour of the original background.
- After the disappearance of the coloured background, an afterimage of the background is induced. This induced colour has a complementary colour to that of the original background. It is possible that this background afterimage induces simultaneous contrast on the "empty shape". Simultaneous contrast is a psychophysical phenomenon of the change in the appearance of a colour (or an achromatic stimulus) caused by the presence of a surrounding average colour (or luminance).

The perceived empty shape effect could also be derived due to both mechanisms.

Medical Conditions

In a visual disturbance called palinopsia, patients have an increased propensity for seeing afterimages, both as a reduced amount of time required to form an afterimage, and an increased duration of the afterimage. Positive afterimages are particularly noticeable, such that even routine eye movement is often accompanied by flickers of what the eye has scanned over the hill (called "tracers"). However, increased negative afterimages are also experienced by palinopsia sufferers. It is unknown if the negative afterimages encountered in palinopsia are formed by then same process described above, although what little research that exists regarding the phenomena suggests that it is brain-related, and not eye-related. Palinopsia can be a persistent condition, but it is also experienced periodically by migraine sufferers.

Ambiguous Image

Ambiguous images are optical illusion images which are crafted to exploit graphical similarities and other properties of visual system interpretation between two or more distinct image forms. These are famous for inducing the phenomenon of multistable perception. Some often used synonymous terms are reversal images, puzzle images and perceptual rivalry.

Ames Room

An Ames room is constructed so that from the front it appears to be an ordinary cubic-shaped room, with a back wall and two side walls parallel to each other and perpendicular to the horizontally level floor and ceiling. However, this is a trick of perspective and the true shape of the room is trapezoidal: the walls are slanted and the ceiling and floor are at an incline, and the right corner is much closer to the front-positioned observer than the left corner (or vice versa). As a result of the optical illusion, a person standing in one corner appears to the observer to be a giant, while a person standing in the other corner appears to be a dwarf. The illusion is so convincing that a person walking back and forth from the left corner to the right corner appears to grow or shrink. Studies have shown that the illusion can be created without using walls and a ceiling; it is sufficient to create an apparent horizon (which in reality will not be horizontal) against an appropriate background, and the eye relies on the apparent relative height of an object above that horizon.

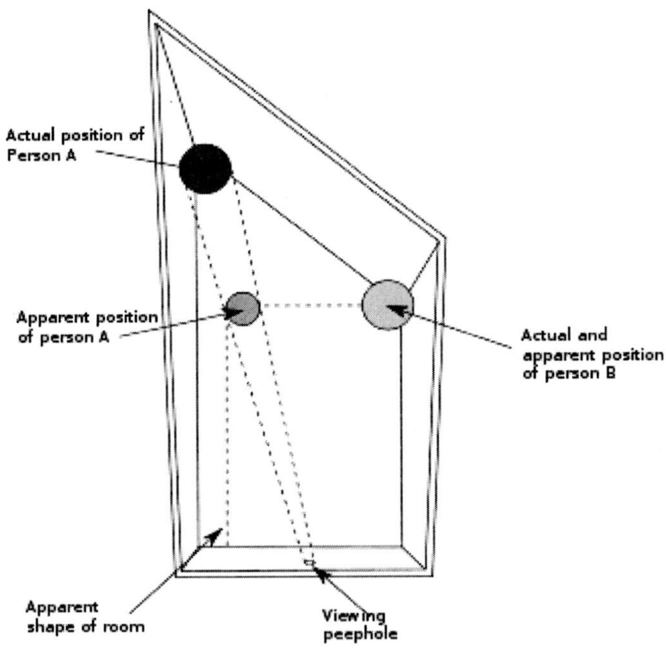

Figure: An Ames room is a distorted room that is used to create an optical illusion. Probably influenced by the writings of Hermann Helmholtz, it was invented by American ophthalmologist Adelbert Ames, Jr. in 1934, and constructed in the following year.

"Anti-gravity" Illusion and "Magnetic Mountains"

Ames' original design also contained a groove that was positioned such that a ball in it appears to roll uphill, against gravity. Richard Gregory regards this apparent "anti-gravity" effect as more amazing than the apparent size changes, although today it is often not shown when an Ames room is exhibited.

He speculates that "magnetic" hills or "gravity" hills can be explained by this principle. For a magic mountain at an unnamed location in Scotland, he found that row of trees form a background similar to the setting of an Ames room, making the water in a creek appear to flow uphill.

For Gregory, this observation raises particularly interesting questions about how different principles for understanding the world compete in our perception. The "anti-gravity effect" is a much stronger paradox than the "size change" effect, because it seems to negate the law of gravity which is a very fundamental feature of the world. It seems counterintuitive that the expectation "rooms are rectangular" can override such a fundamental experience. In contrast, the apparent size change is not such a strong paradox, as we do have the experience that objects can change size to a certain degree (for example, people and animals can become smaller or larger by crouching or stretching).

Honi Phenomenon

A type of selective perceptual distortion known as the Honi phenomenon causes some married persons to perceive less size distortion of the spouse than a stranger in an Ames room.

The effect was related to the strength of love, liking, and trust of the spouse being viewed. Women who were high positive in this area perceived strangers as being more distorted than their partners. Size judgements by men did not seem to be influenced by the strength of their feeling toward their spouse.

Further study has come to the conclusion that the Honi phenomenon does not reliably exist as first thought, but may be explained as sex difference influencing perception, with women interpreting a larger reading as a more meaningful or valuable perception of things than men's.

In Media

The Ames room principle has been used widely in TV and movie productions for special effects when it was necessary to show actors in giant size next to actors in small size. For example, production of "The Lord of the Rings film trilogy used several Ames room sets in Shire sequences to make the heights of the diminutively-sized hobbits correct when standing next to the taller Gandalf.

When used for special effects, the viewers will not see that an Ames room is being used. However, a few times an Ames room has also been shown explicitly. An Ames room is depicted in the 1971 film adaptation of the Roald Dahl novel Charlie and the Chocolate Factory. The 1960s television series Voyage to the Bottom of the Sea used an Ames room in one episode to show, rather than just declare, an attempt to make two characters (one standing on each side of the room) lose their minds. The video for "The Denial Twist" by The White Stripes employs an Ames room and other techniques of false perspective.

The 2010 HBO film Temple Grandin utilised an Ames room in the opening title sequence, and later figures in the actual story, where the title character, with high-functioning autism, intuitively devises the illusion in a scale model as a science project. The cover of "On the Level", a 1975 album by the rock band Status Quo features the band in an Ames room.

Ames Trapezoid

The Ames trapezoid or Ames window is a style of window which, when observed frontally, appears to be a rectangular window but is, in fact, a trapezoid. The window is mounted on a rod connected to an electric motor that rotates it about its vertical axis. When it is observed with one eye from about 3 metres or with both eyes at 6 metres, or more the window appears to rotate through 180 degrees and then seems to stop momentarily and reverse its direction of rotation. It is therefore not perceived vertically to be rotating continuously in one direction but instead is mis-perceived to be oscillating, reversing its direction once every 180 degrees.

3

Forming of Images

Impossible Object

An impossible object (also known as an impossible figure or an undecidable figure) is a type of optical illusion consisting of a two-dimensional figure which is instantly and subconsciously interpreted by the visual system as representing a projection of a three-dimensional object although it is not actually possible for such an object to exist (at least not in the form interpreted by the visual system). In most cases the impossibility becomes apparent after viewing the figure for a few seconds. However, the initial impression of a 3D object remains even after it has been contradicted. There are also more subtle examples of impossible objects where the impossibility does not become apparent spontaneously and it is necessary to consciously examine the geometry of the implied object to determine that it is impossible. *Impossible objects* are of interest to psychologists, mathematicians and artists without falling entirely into any one discipline.

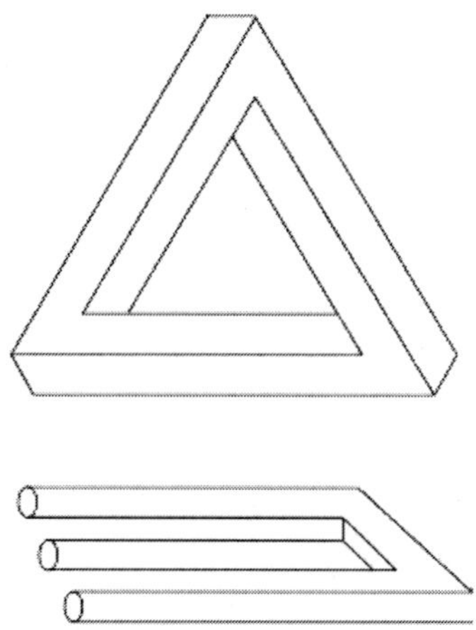

Figure: Two famous undecidable figures, the Penrose triangle and devil's tuning fork.

Notable Examples

Figure: Viewed from a certain angle, thiscube appears to defy the laws of geometry.

Notable undecidable figures include:

- Necker Cube
- Penrose stairs
- Penrose triangle
- Blivet (or Devil's tuning fork).

History

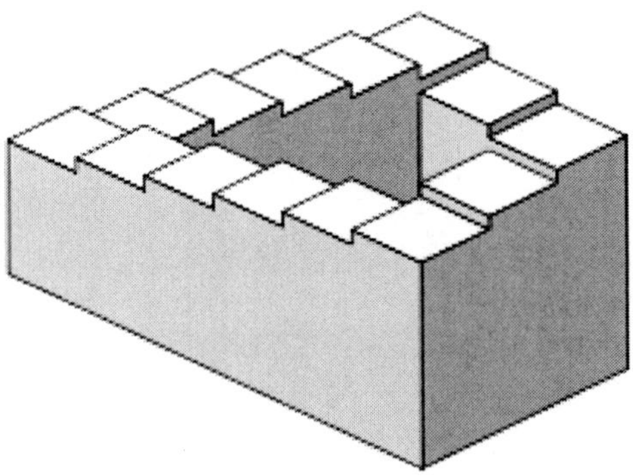

Figure: The Penrose stairs

Swedish artist Oscar Reutersvärd was the first to deliberately design many impossible objects. He has been called "the father of impossible figures". In 1934 he drew the Penrose triangle, some years before the Penroses. In Reutersvärd's version the sides of the triangle are broken up into cubes. In 1956, British psychiatrist Lionel Penrose and his son, mathematician Roger Penrose, submitted a short article to the British Journal of Psychology titled *Impossible Objects: A Special Type of Visual Illusion*. This was illustrated with the Penrose Triangle and Penrose stairs. The article referred to Escher, whose work had sparked their interest in the subject, but not Reutersvärd, whom they were unaware of. The article was only published in 1958.

From the 1930s onwards Dutch artist M. C. Escher produced many drawings featuring paradoxes of perspective gradually working towards impossible objects. In 1957 he produced his first drawing containing a true impossible object: *Cube with Magic Ribbons*. He produced many further drawings featuring impossible objects, sometimes with the entire drawing being an undecidable figure. His work did much to draw the attention of the public to impossible objects. Some contemporary artists are also experimenting with impossible figures, for example, Jos de Mey, Shigeo Fukuda, Sandro del Prete, Istvan Orosz (Utisz), Guido Moretti, Tamas F. Farkas and Mathieu Hamaekers.

In Fiction

- In H. P. Lovecraft's celebrated short story, *The Call of Cthulhu* (1928), the titular entity emerges from the vast and ancient alien city of R'lyeh, which is described in terms of "non-Euclidean geometry" and contains angles which are "all wrong" (appearing acute but behaving as if obtuse, for example) and planes which could be horizontal or slanted depending on how the observer looks at them. The overall description of 'impossible geometry' seems to strongly prefigure Escher's work.
- In the 1982 *Doctor Who* story *Castrovalva*, the titular town is subject to apparently impossible architecture. The storyline also relied heavily on recursion, a favourite theme in Escher's most famous works, and used ideas taken from *Belvedere*, *Ascending and Descending*, and *Relativity*. Despite the use of its name, the Escher lithograph of the same name does not feature impossible architecture.
- In the *Star Trek: The Next Generation* episode "I, Borg", a plan was made to destroy the entire race of Borg—malevolent cybernetic aliens whose minds were interconnected—by showing one of the Borg a picture of a highly complex impossible object. This image would be transmitted back to the Borg hive, overloading

its consciousness in larger and larger attempts to understand the image. This plan was dismissed as being genocide, so its potential results were never seen.

- In the computer game *Diablo II*, parts of the "Arcane Sanctuary" region were based on impossible drawings.
- Alan Moore's miniseries *1963* features a character called the Hypernaut that lives in a space station shaped like an impossible object.
- In the video game *Eternal Darkness: Sanity's Requiem* the Essence of Xel'lotath, the Sigil of Xel'lotath, is an impossible artifact resembling a warped angel.
- In *The Simpsons* episode "Treehouse of Horror IV", homer is seen chasing bart in a painting parodying the famous M.C. Escer painting "Ascending and descending" "Treehouse of Horror VIII", a blivet (or devil's tuning fork) can be seen on Professor Frink's yard sale. In *The Simpsons Movie*, Homer falls down miles of the impossible infinite House of Stairs while having an epiphany. The couch gag in the episode "The PTA Disbands", the family is running through an Escher-like living room.
- On the cover of July 1985 Mad magazine, a picture describing devil's tuning fork can be seen upon the wall along with a stuffed St. Bernard's head on the wall of Alfred E. Neuman's living room
- The video game *Echochrome* for the PlayStation Portable and PlayStation 3 features a mannequin who must traverse a series of impossible objects.
- The set of the climactic scene in the Goblin King's castle in the 1986 fantasy film, *Labyrinth*, directed by Jim Henson, is based on M. C. Escher's "Relativity", and features impossible staircases and perspectives.
- In the big-screen film version of "The Avengers", Mrs Peel (Uma Thurman) is trapped in a labyrinthine mansion which features a direct downward view of a Penrose Stair.
- In *Pokémon Platinum* there is an area called "Distortion World" full of impossibilities such as these.
- In the *Family Guy* episode "No Meals on Wheels" there is a scene parodying students propensity to poster dorm rooms with "crazy stairs" a.k.a. M. C. Escher's *Relativity*.

- In "Science Girls", made by Hanako Games, the wormhole that leads to the aliens' home planet has a double Penrose Triangle as its power source, possibly to create a spacial paradox to warp dimensions.
- In the video game "Prince of Persia 2: The Shadow and the Flame" the final battle features impossible geometry when you jump down the vertical part of an impossible triangle
- The 1984 game *Realm of Impossibility* features levels which largely involve impossible objects.
- In the 2010 movie *Inception*, Arthur shows Ariadne how to install a Penrose staircase in dreams. He later uses the same configuration to flank a pursuing enemy.
- The first episode of the second season of Xiaolin Showdown has a fight sequence in which the protagonist chases his opponent through a building constructed of various Escheresque rooms.
- There is also significant reference to an impossible object in Jonathan Lethem's most recent novel, "Chronic City." Where it plays a major part in the storyline.

Isometric Illusion

Figure: A photograph of a three-dimensional installation showing the logo of science centre Heureka. The logo can be viewed in two different ways; with the cubes appearing to be pushed outward, or, as it in reality is, the cubes pushed inward.

An isometric illusion (also called an ambiguous figure or inside/outside illusion) is a type of optical illusion, specifically one due to multistable perception.

In general, any shape built entirely of same-length (i.e., *isometric*) lines that does not clearly indicate relative direction between its components will evoke such a perceptual "flip-flopping". The Necker Cube is a famous example of an isometric illusion.

Jastrow Illusion

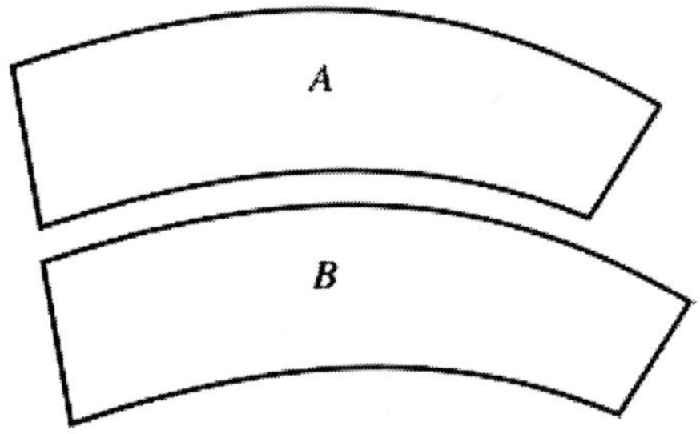

Figure: *Jastrow illusion*

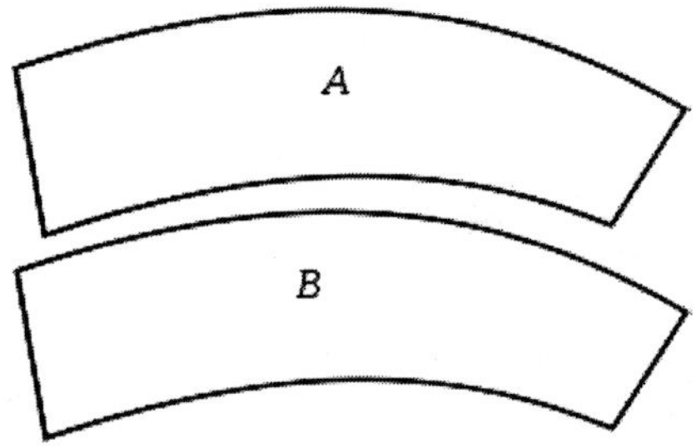

Figure: The two figures are identical.

The Jastrow illusion is an optical illusion discovered by the American psychologist Joseph Jastrow in 1889. In this illustration, the two figures are identical, although the lower one appears to be larger. The short edge of the upper figure is compared to the long side of the lower one.

Leaning Tower Illusion

The Leaning Tower Illusion is a visual illusion that presents two identical images of the Leaning Tower of Pisa side by side. Although the images are duplicates, one has the impression that the tower on the right leans more, as if photographed from a different angle. The illusion was discovered by Frederick Kingdom, Ali Yoonessi and Elena Gheorghiu at McGill University, and won first prize in the Best Visual Illusion of the Year Contest 2007.

The authors suggest that the illusion occurs because of the way the visual system takes into account perspective. When two identical towers rise in parallel but are viewed from below, their corresponding outlines converge in the retinal image due to perspective. The visual system 'corrects' for the perspective distortion and sees the towers as they are, i.e. rising in parallel. However in the two identical images of the Pisa tower, the corresponding outlines do not converge but are parallel, and are hence assumed to arise from towers that are non-parallel, i.e. diverging. The illusion reveals that the visual system is obliged to treat the two images as part of the same scene, in other words see them as the 'Twin Towers of Pisa'.

Although the Pisa tower demonstrates the illusion and provides a pun for its name, the illusion can be seen in any pair of (identical) images of a receding object.

Lilac Chaser

Lilac chaser is a visual illusion, also known as the Pac-Man illusion. It consists of 12 lilac (orpink, rose or magenta), blurred discs arranged in a circle (like the numbers on a clock), around a small black, central cross on a grey background. One of the discs disappears briefly (for about 0.1 second), then the next (about 0.125 second later), and the next, and so on, in a clockwise direction. When one stares at the cross for about 20 seconds or so, one sees three different things:

1. A gap running around the circle of lilac discs;
2. A green disc running around the circle of lilac discs in place of the gap;
3. The green disc running around on the grey background, with the lilac discs having disappeared in sequence.

The illusion is due to three, well-known effects:

1. An illusion of movement called beta movement;
2. An afterimage in which an opposite, complementary, colour—green—appears when each lilac spot disappears (if the discs were blue, one would see yellow);
3. Troxler's fading of the lilac discs.

History

The illusion was invented by Jeremy Hinton some time before 2005. He stumbled across the configuration while devising stimuli for visual motion experiments. In one version of a program to move a disc around a central point, he mistakenly neglected to erase the preceding disc, which created the appearance of a moving gap. On noticing the moving green-disc afterimage, he adjusted foreground and background colours, number of discs, and timing to optimise the effect.

In 2005 Hinton blurred the discs, allowing them to disappear when a viewer looks steadily at the central cross. Hinton entered the illusion in the ECVP Visual Illusion Contest, but was disqualified for not being registered for that year's conference. Hinton approached

Michael Bach, who placed an animated GIF of the illusion on his web page of illusions, naming it the "Lilac Chaser", and later presenting a configurable Java version. The illusion became popular on the Internet in 2005.

Explanation

The lilac chaser illusion combines three simple, well-known effects:

When a visual event occurs briefly at one place in the visual field, and then a similar event occurs at an adjacent place in the same visual field, we perceive movement from the first place to the second. This is called apparent movement or beta movement, because no actual movement has occurred. Beta movement is the basis of moving neon signs, film, and video. We see movement because such displays stimulate receptors (called Reichardt detectors) in ourbrains that encode movement. The visual events in lilac chaser initially are the disappearances of the lilac discs. The visual events then become the appearances of green afterimages.

When a lilac stimulus that is presented to a particular region of the visual field for a long time (say 10 seconds or so) disappears, a green afterimage will appear. The afterimage lasts only a short time, and in this case is effaced by the reappearance of the lilac stimulus. The afterimage is a simple consequence of adaptation of the rods and cones of the retina. Colour and brightness are encoded by the ratios of activities in three types of cones (and also the rods under mesopic conditions).

The cones stimulated by lilac get "tired". When the stimulus disappears, the tiredness of some of the cones means that the ratios evoked by the grey background are the same as if a green stimulus had been presented to these cones when they are fresh. Adaptation of rods and cones begins immediately when they are stimulated, so afterimages also start to grow. We normally do not notice them because we move our eyes about three times a second, so the image of a stimulus constantly falls on new, fresh, unadapted rods and cones. In lilac chaser, we keep our eyes still, so the afterimages grow and are revealed when the stimulus disappears.

When a blurry stimulus is presented to a region of the visual field away from where we are fixating, and we keep our eyes still, that stimulus will disappear even though it is still physically presented. This is called Troxler's fading. It occurs because although our eyes move a little when we are fixating a point, away from that point (in *peripheral vision*) the movements are not large enough to shift the lilac discs to new neurons of the visual system. Their afterimages essentially cancel the original images, so that all one sees of the lilac discs is grey, except for the gap where the green afterimage appears.

These effects combine to yield the remarkable sight of a green spot running around in a circle on a grey background when only stationary, flashing lilac spots have been presented. Occasionally it seems as though the green afterimage has eaten up the lilac discs, this resemblance's to Pac-Man accounting for the illusion's alternative name.

Psychophysics

As of May 2011, no systematic study of the stimulus properties of the illusion had been published. Hinton optimised the conditions for all three aspects of the illusion before releasing it. He also noted that the colour of the green disc could be outside the colour gamut of the monitor on which it was created (because the monitor never displays the green disc, only lilac ones). Michael Bach's version of the illusion allows viewers to adjust some aspects of the illusion. It is simple to confirm that the illusion occurs with other colours, and that Troxler fading is enhanced by reducing the saturation of the discs.

Other Effects

The demonstration in this article shows a blurred, very slightly blue-grey disc in each gap. These can be seen by examining one frame of the image under high magnification with any adjoining lilac discs masked off. But these blurred discs are not required for the illusion. Michael Bach's demonstration contains no such discs, but yields the illusion. It is not necessary to fixate the black cross for the effects to occur. As long as the eyes are held steadily on any point of the figure, even the centre of one of the discs, the illusion will occur.

If instead of fixating the black cross, one follows the moving gap with one's eyes, one will see only a moving gap and 12 lilac discs rather than a single green disc. This is because the green disc arises as an afterimage, requiring the eyes to be held steadily to occur, and the disappearance of the lilac discs is from Troxler's fading, also requiring the eyes to be held steadily to occur. Moving the eyes prevents the formation of afterimages and the occurrence of Troxler's fading.

If after looking at the effect for 30 seconds or so, one moves one's eyes elsewhere (e.g., to another point on the figure or to a blank sheet of white paper), one will see a stationary ring of 12 green discs that will fade after a short time. These green discs are the afterimages of the 12 lilac discs. If one watches the illusion for long enough to see only the green disc and then moves away from the computer screen while keeping the eyes on the cross, one sees larger green spots outside a ring of lilac spots with a smaller green disc running around them. The smaller green disc may merge briefly with the outer green spots, making the spots appear to be radial blobs. The outer green spots soon fade. These outer green spots are afterimages that appear larger because of Emmert's law: the size of an afterimage becomes larger as its viewing distance is increased.

They are outside because moving away from the computer screen has decreased the visual angle of the lilac spots. They fade because the lilac discs that constantly refresh the green afterimages are now projected onto a different part of the retina. If one moves towards the screen, the effects are opposite.

If one closes the right eye and moves close to the stimulus so that the nine-o'clock disc falls in the blind spot, one sees that the movement is no longer smooth. There is a noticeable pause when the disappearance of the disc occurs on the region of the retina having no rods or cones. This suggests there are limits to the filling in that normally prevents us from noticing a black hole in our visual fields at the location of the blind spot.

Liquid Crystal Shutter Glasses

Liquid crystal shutter glasses (also called LC shutter glasses or active shutter glasses.) are glasses used in conjunction with a display screen to create the illusion of a three dimensional image, an example of stereoscopy. Each eye's glass contains a liquid crystal layer which has the property of becoming dark when voltage is applied, being otherwise transparent. The glasses are controlled by an infrared, radio frequency, DLP-Link or Bluetooth transmitter that sends a timing signal that allows the glasses to alternately darken over one eye, and then the other, in synchronisation with the refresh rate of the screen. Meanwhile, the display alternately displays different perspectives for each eye, using a technique called alternate-frame sequencing, which achieves the desired effect of each eye seeing only the image intended for it.

Advantages

LC shutter glasses mostly eliminate "ghosting" which is a problem with other 3D display technologies such as linearly polarised glasses. Moreover, unlike red/cyan colour filter 3D glasses, LC shutter glasses are colour neutral enabling 3D viewing in the full colour spectrum.

Disadvantages

Flicker can be noticeable except at very high refresh rates, as each eye is effectively receiving only half of the monitor's actual refresh rate. Most modern LC glasses however generally work in higher refresh rates and mostly eliminate this problem.

Until recently, the method only worked with CRT monitors; some modern flat-panel monitors now support high-enough refresh rates to work with some LC shutter systems.

LC shutter glasses are shutting out light half of the time; moreover, they are slightly dark even when letting light through, because they are polarised. This gives an effect similar to watching TV with sunglasses on, which causes a darker picture perceived by the viewer.

However, this effect can produce a higher perceived display contrast when paired with LCD displays because of the reduction in backlight bleed. Frame rate has to be double that of an ordinary stream to get an equivalent result. All equipment in the chain has to be able to process frames at double rate; in essence this doubles the hardware requirements of the equipment. This is especially noticeable when the image stream is interactively generated in real time by 3D hardware on computers.

In addition, shutter glasses tend to be much more expensive than other forms of stereoscopic glasses. Whereas most anaglyph, ChromaDepth, and polarised 3D glasses can be purchased at very low prices (less than US$1 as of 2010, with anaglyph filters being the least expensive), shutter glasses feature far more advanced technology and usually sell for two orders of magnitude higher than paper anaglyphs and one order over paper ChromaDepth and polarised glasses, with most models selling for well over US$100, particularly for the standard wireless models.

Shutter glasses are also matched to the TV so it may not be possible to use your shutter glasses with a different brand 3DTV. However, efforts are being made to create a Universal 3D Shutter Glass. In March 2011 Panasonic Corporation, together with XPAND 3D, have formulated the M-3DI Standard, which aims to provide industry-wide compatibility and standardisation of LC (Active) Shutter Glasses. This movement aims to bring about compatibility among manufacturers of 3D TV, computer, notebook, home projection, and cinema with standardised LC (Active) Shutter Glasses that will work across all 3D hardware seamlessly.

LC Glasses Providers

There are many sources of low-cost 3D glasses. IO glasses are the most common glasses in this category. XpanD 3D is a manufacturer of shutter glasses, with over 1000 cinemas currently using XpanD glasses. With the release of this technology to the home-viewer market as of 2009, many other manufacturers are now developing their own LC shutter glasses, such as Unipolar International Limited, Accupix Co., Ltd, Panasonic, Tnitrend, Samsung, and Sony. The M-3DI Standard, announced by Panasonic Corporation together with XPAND 3D in March 2011, aims to provide industry-wide compatibility and standardisation of LC (Active) Shutter Glasses.

Samsung has developed Active 3D glasses that are 2 ounces and utilise lens and frame technology pioneered by Silhouette who creates glasses for NASA. Nvidia makes a 3D Vision kit for the PC; it comes with 3D shutter glasses, a transmitter, and special graphics driver software. A certified 120 Hz monitor is required to use the 3D Vision; standard LCD monitors run at 60 Hz.

History

Liquid crystal shutter glasses were first invented by Stephen McAllister of Evans and Sutherland Computer Corporation in the mid-1970s. The prototype had the LCDs mounted to a small cardboard box using duct tape. The glasses were never commercialised due to ghosting, but E&S was a very early adopter of third-party glasses such as the StereoGraphics CrystalEyes in the mid-1980s.

Compatible Hardware

CRT Display

In 1982, Sega's arcade video game *SubRoc-3D* came with a special 3-D eyepiece, which was a viewer with spinning discs to alternate left and right images to the player's eye from a single monitor. In 1984, the Vectrex released the 3D Imager, a primitive form of active shutter glasses that used a motorized rotating disc with transparencies as physical shutters. Although bulky and crude, they used the same basic principle of rapidly alternating imagery that modern active shutter glasses still use.

Sega released the SegaScope 3-D for the Sega Master System in 1987 for Japan and in 1989 for North America, where it was the first known electronic device to use LCD Active Shutter glasses. Only eight 3D compatible games were ever released. Nintendo's Famicom also featured a similar Famicom 3D System, which was an LCD shutter headset, released in 1987 only for Japan. In 1999-2000, a number of companies created stereoscopic LC shutter glasses kits for the Windows PCs which worked with application and games written for Direct3D andOpenGL 3D graphics APIs. These kits only worked with CRT computer displays and employed either VGA pass-through, VESA Stereo or proprietary interface for left-right synchronisation.

The most prominent example was the ELSA Revelator glasses, which worked exclusively in NVidia cards through a proprietary interface based on VESA Stereo. Nvidia later bought the technology and used it in its stereo driver for Windows. The glasses kits came with driver software which intercepted API calls and effectively rendering the two views in sequence; this technique required twice the performance from the graphic card, so a high-end device was needed. Visual glitches were common, as many 3D game engines relied on 2D effects which were rendered at the incorrect depth, causing disorientation for the viewer. Very few CRT displays were able to support 120 Hz refresh rate at common gaming resolutions of the time, so high-end CRT display was required for flicker-free image; and even with capable CRT monitor, many users reported flickering and headaches.

These CRT kits were entirely incompatible with common LCD monitors which had very high pixel response times, unlike CRT displays. Moreover, the display market swiftly shifted to LCD monitors and most display makers ceased production of CRT monitors in early 2000s, which meant that PC glasses kits shortly felt into disuse and were reduced to a very niche market, requiring a purchase of a used high-end, big diagonal CRT monitor.

SplitFish Eye FX 3D was a stereo 3D shutter glasses kit for the Sony PlayStation 2 released in 2005; it only supported standard-definition CRT TVs. The accessory included a pass-trough cable for the PS2 gamepad; when activated, the attached accessory would issue a sequence of rapidly alternating left-right movement commands to the console, producing a kind of "wiggle stereoscopy" effect additionally aided by the wired LC shutter glasses which worked in sync with these movements. The kit arrived too late in the product cycle of the console when it was effectively replaced by the PlayStation 3, and only a few games were supported, so it was largely ignored by gamers.

The USB-based Nvidia 3D Vision kit released in 2008 supports CRT monitors capable of 100, 110, or 120 Hz refresh rate, as well as 120 Hz LCD monitors.

3D DLP TV

In 2007, Texas Instruments introduced stereo 3D capable DLP solutions to its OEMs, and Samsung and Mitsubishi introduced the first 3D ready televisions. These solutions utilise the inherent speed advantage of the Digital Micro-mirror Device (DMD) to sequentially generate a high refresh rate for the left and right views required for stereoscopic imaging.

DLP 3D technology uses the SmoothPicture wobulation algorithm and relies on the properties of modern 1080p60 DMD imagers. It effectively compacts two L/R views into a single frame by using a checkerboard pattern, only requiring a standard 1080p60 resolution for stereoscopic transmission to the TV. The claimed advantage of this solution is increased spatial resolution, unlike other methods which cut vertical or horizontal resolution in half.

The micromirrors are organised in a so-called "offset-diamond pixel layout" of 960×1080 micromirrors, rotated 45 degrees, with their centre points placed in the centre of "black" squares on the checkerboard.

The DMD employs full-pixel wobulation to display the complete 1080p image as two half-resolution images in a fast sequence. The DMD operates at twice the refresh rate, i.e. 120 Hz, and the complete 1080p picture is displayed in two steps. On the first cadence, only half of the original 1080p60 image is displayed - the pixels that correspond to the "black" squares of the checkerboard pattern. On the second cadence, the DMD array is mechanically shifted (wobulated) by one pixel, so the micromirrors are now in a position previously occupied by the gaps, and another half of the image is displayed - this time, the pixels that correspond to the "white" squares.

A synchronisation signal is generated for LC shutter glasses worn by the viewer, using either a standard VESA Stereo plug to connect wired glasses or wireless emitters, or brief flashes of light on the viewing screen during the blanking interval (DLP Link). The LC shutter glasses process the signal and control the shutter for each eye to ensure that the correct left and right views are presented to the correct eye.

3D PDP

Plasma display panels are inherently high-speed devices as well, since they use pulse-width modulation to maintain the brightness of individual pixels, making them compatible with sequential method involving shutter glasses. Modern panels feature pixel driving frequency of up to 600 Hz and allow 10 bit to 12 bit colour precision with 1024 to 4096 gradations of brightness for each subpixel. Samsung Electronics launched 3D ready PDP TVs in 2008, a "PAVV Cannes 450" in Korea and PNAx450 in the UK and the US. The sets utilise the same checkerboard pattern compression scheme as their DLP TVs, though only at the native resolution of 1360x768 pixels and not at HDTV standard 720p, making them only usable with a PC.

Matsushita Electric (Panasonic) prototyped the "3D Full-HD Plasma Theatre System" on CES 2008. The system is a combination of a 103-inch PDP TV, a Blu-ray Disc player and shutter glasses. The new system transmits 1080i60 interlaced images for both right and left eyes, and the video is stored on 50-Gbyte Blu-ray using the MPEG-4 AVC/H.264 compression Multiview Video Coding extension.

LCD/LED TV

Liquid crystal displays have traditionally been slow to change from one polarisation state to another. Users of early 1990s laptops are familiar with the smearing and blurring that occurs when something moves too fast for the LCD to keep up. This smearing can result in a completely unviewable image when using shutter glasses.

LCD technology is not usually rated by frames per second but rather the time it takes to transition from darkness to brightness and back to darkness, in milliseconds. In order to achieve an equivalent minimum refresh rate of 120 Hz, an LCD must be able to transition at a speed of not more than 8.33 ms. However, each frame is displayed for at most 8.33ms, and minimising the response is key. For example: if it takes 8.33ms for the LCD to transition to the desired image, and a sequential black/white image is shown, the 8.33ms which should be displaying "white" will begin at black, and after 8.33ms finally achieve white. Similarly, the next 8.33ms which should be displaying "black" will begin as white, and after 8.33ms finally achieve black.

However, because pixel transition speed has become a strong selling point of LCD monitors, marketing hype has unfortunately obscured these speed-of-transition specifications with what some consider tortuous qualifying statements that make inadequate technology appear to be better than it really is. While the average person attempting to buy a high quality LCD for normal home use might not notice these minor performance differences, a slowly transitioning LCD can have a severely negative impact on usability with shutter glasses. For stereoscopic applications, it is important that the LCD be truly capable of what is being claimed.

Mach Bands

Mach bands is an optical illusion named after the physicist Ernst Mach. The illusion consists of light or dark stripes that are perceived next to the boundary between two regions of an image that have different lightness gradients (even if the lightness itself is the same on both sides of the boundary).

Explanation

The Mach bands effect is due to the spatial high-boost filtering performed by the human visual system on the luminance channel of the image captured by the retina. This filtering is largely performed in the retina itself, by lateral inhibition among its neurons. The effect is independent of the orientation of the boundary.

Mach Bands in Radiology

This visual phenomenon is important to keep in mind when evaluating dental radiographs for evidence of decay, in which grayscale images of teeth and bone are analysed for abnormal variances of intensity. Afalse-positive radiological diagnosis of dental cariescan easily arise if the practitioner doesn't take into account the likelihood of this illusion. Not only do Mach bands manifest adjacent to metal restorations or appliances, as in the radiograph at right, but they can also present at the boundary between enamel and dentin. Mach bands may also result in the misdiagnosis of horizontal root fractures because of the differing radiographic intensities of tooth and bone.

Magic Eye

Magic Eye is a series of books published by N.E. Thing Enterprises (renamed in 1996 to Magic Eye Inc.). The books feature autostereograms (precisely, random dot autostereogram), which allow people to see 3D images by focusing on 2D patterns. The viewer must diverge his or her eyes in order to see a hidden three-dimensional image within the pattern. "Magic Eye" has become something of agenericized trademark, often used to refer to autostereograms of any origin. The autostereogram predates the Magic Eye series by several years. Christopher Tyler created the first autostereograms in 1979.

Unable to find an American publisher after creating its first images in 1991, creators Tom Baccei and Cheri Smith managed to make a deal with Tenyo, a Japanese company that sells magic supplies. Tenyo published its first book in late 1991 titled *Miru Miru Mega Yokunaru Magic Eye* ("Your Eyesight Gets Better & Better in a Very Short Rate of Time: Magic Eye"), sending sales representatives out to street corners to demonstrate how to see the hidden image. Within a few weeks the first Japanese book became a best seller, as did the second, rushed out shortly after. The first North American Magic Eye book, *Magic Eye: A New Way of Looking at the World* was released by Andrews & McMeel in 1993.

Within a year it had been followed by two sequels that were also extremely popular. The three books spent a cumulative total of 73 weeks on the New York Times Bestseller List. Dozens of other books and other products have been released since then; Magic Eyestereograms have been featured on postcards, mousepads, lunch boxes, and even neckties. Magic Eye stereograms have also been featured in a weekly newspaper comic panel syndicated by Universal Press Syndicate.

The Magic Eye images have a horizontally repeating pattern which differs slightly with each repetition, therefore giving the illusion of depth when each eye focuses on a different part of the pattern. The stereograms were created using a patented process that allows colourful patterns to be used in creating the final images. Magic Eye stereograms have been used by orthoptists and vision therapists in the treatment of some binocular vision and accommodative disorders.

McCollough Effect

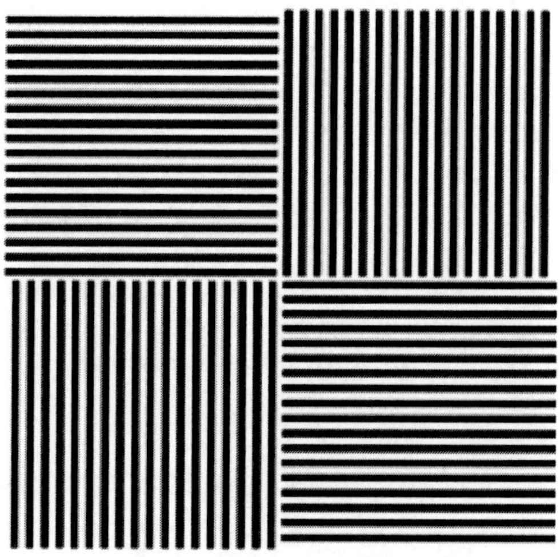

Figure: A test image for the McCollough effect. On first looking at this image, the vertical and horizontal lines should look black and white, colourless. After induction, the horizontal lines should look greenish and the vertical lines should look pinkish.

Figure: One induction image for the McCollough effect. Stare at the centre of this image for a few seconds, then at the centre of the second image below for a few seconds. Then return to this image. Keep looking between the two coloured images for at least 3 minutes.

Figure: A second induction image for the McCollough effect. Stare at the centre of this image for a few seconds, then at the centre of the image above for a few seconds. Then return to this image. Keep looking between the two coloured images for at least 3 minutes.

The McCollough effect is a phenomenon of human visual perception in which colourless gratings appear coloured depending on (*contingent* on) the orientation of the gratings. It is an aftereffect requiring a period of induction to produce it. For example, if someone alternately looks at a red horizontal grating and a green vertical grating for a few minutes, a black-and-white horizontal grating will then look greenish and a black-and-white vertical grating will then look pinkish. The effect was discovered by Celeste McCollough in 1965.

Producing the Effect

To obtain the effect, first look at a test image similar to that at top right. It should contain oppositely oriented gratings of lines, such as horizontal and vertical as shown here. Next, stare alternately at two induction images similar to the ones directly beneath the top image. One image should show one orientation of grating (here horizontal) with a coloured background (red) and the other should show the other orientation of grating (here vertical) with a different, preferablyoppositely coloured background (green). Each image should be gazed at for several seconds at a time, and the two images should be gazed at for a total of several minutes for the effect to become visible. Stare approximately at the centre of each image, allowing the eyes to move around a little. After several minutes, look back to the test image; the gratings should appear tinted by the opposite colour to that of the induction gratings (i.e., horizontal should appear greenish and vertical pinkish).

Properties of the Effect

The McCollough effect is remarkable because it is long lasting. McCollough originally reported that these aftereffects may last for an hour or more. They can last much longer than that, however. Jones and Holding (1975) found that 15 minutes of induction can lead to an effect lasting 3 months.

The effect is different from coloured afterimages, which appear superimposed on whatever is seen and which are quite brief. It depends on retinal orientation (tilting the head by 45 degrees makes the colours in the above example disappear; tilting the head by 90 degrees makes the colours reappear such that the gravitationally vertical grating now looks green), and because inducing the effect with one eye leads to no effect being seen with the other eye. However, there is some evidence of binocular interactions.

Any aftereffect requires a period of induction (or adaptation) with an induction stimulus (or, in the case of the McCollough effect, induction stimuli). It then requires a test stimulus on which the aftereffect can be seen.

In the McCollough effect as described above, the induction stimuli are the red horizontal grating and the green vertical grating. A typical test stimulus might show adjacent patches of black-and-white vertical and horizontal gratings (as above). The McCollough-effect colours are less saturated than the induction colours.

The induction stimuli can have any different colours. The effect is strongest, however, when the colours are complementary, such as red and green, and blue and orange. A related version of the McCollough effect also occurs with a single colour and orientation.

For example, induction with only a red horizontal grating makes a black-and-white horizontal test grating appear greenish whereas a black-and-white vertical test grating appears colourless (although there is some argument about that). Stromeyer (1978) called these non-redundant effects. According to him, the classic effect with induction from two different orientations and colours simply makes the illusory colours more noticeable via contrast. The effect is specific to the region of the retina that is exposed to the induction stimuli. This has been shown by inducing opposite effects in adjacent regions of the retina (i.e., from one region of the retina verticals appear pink and horizontals appear greenish; from an adjacent region of the retina, verticals appear greenish and horizontals appear pink).

Nevertheless, if a small region of the retina is exposed to the induction stimuli, and the test contours run through this region, the effect spreads along those test contours. Of course, if the induced area is in the fovea (central vision) and the eyes are allowed to move, then the effect will appear everywhere in the visual scene visited by the fovea. The effect is also optimal when the thickness of the bars in the induction stimulus matches that of those in the test stimulus (i.e., the effect is tuned, albeit broadly, to spatial frequency). This property led to non-redundant effects being reported by people who had used computer monitors with uniformly coloured phosphors to do word processing. These monitors were popular in the 1980s, and commonly showed text as green on black. People noticed later when reading text of the same spatial frequency, in a book say, that it looked pink. Also a horizontal grating of the same spatial frequency as the horizontal lines of the induction text (such as the horizontal stripes on the letters "IBM" on the envelope for early floppy disks) looked pink.

A variety of similar aftereffects have been discovered not only between pattern and colour contingencies, but between movement/colour, spatial frequency/colour and other relationships. All such effects may be referred to as McCollough Effects or MEs.

Explanations of the Effect

McCollough's paper sparked hundreds of scientific papers on the effect. Explanations of the McCollough effect appear to fall in to three camps. McCollough indicated colour adaptation of edge sensitive neurons in lower, monocular regions of the visual cortex.

A functional explanation of MEs has been posited in the form of an error-correcting device (ECD) whose purpose is to maintain an accurate internal representation of the external world. Consistent pairings of colour and oriented lines are not found frequently in natural environments, thus consistent pairing may indicate pathology of the eye. An ECD might compensate for such pathology by adjusting the appropriate neurons to a neutral point in adaptation to orientation contingent colour. A third explanation points to the contribution of classical conditioning to normal homeostatic regulation. MEs are explained by the same mechanisms as pharmacological withdrawal symptoms, thus the "pharmacological CR is expressed as pharmacological adaptation (tolerance) in the presence of the drug, and withdrawal symptoms in the absence of the drug" and the "chromatic CR is expressed as chromatic adaptation in the presence of colour, and the ME in the absence of colour". By this account MEs are of no adaptive value, but have been selected for as a domain-general ability to anticipate events. This is related to Opponent-process theory.

It is worth noting that these theories do not predict or explain the Anti-McCollough Effect.

Neurophysiological explanations of the effect have variously pointed to the adaptation of cells in the lateral geniculate nucleus designed to correct for chromatic aberration of the eye, to adaptation of cells in the visual cortex jointly responsive to colour and orientation (this was McCollough's explanation) such as monocular areas of cortical hypercolumns, to processing within higher centres of the brain (including the frontal lobes), and to learning and memory.

In 2006, the explanation of the effect was still the subject of debate, although there was a consensus in favour of McCollough's original explanation. MEs do not transfer interocularly and from this it seems reasonable to deduce that the effect occurs in an area of the visual system prior to V1-4B, where binocular cells first occur.

The Anti-McCollough Effect

Recently, a new effect in the opposite direction of the McCollough effect was discovered and has been termed the anti-McCollough effect. This effect may be induced by alternating pairings of gratings in parallel alignment, one achromatic (black and white) and the other black and a single colour (say black and red). If the colour used was red, then after the induction phase the achromatic grating appeared slightly red.

This effect is distinct from the classical effect in three important regards; the perceived colour of the aftereffect is the same as the inducer's colour, the perceived colour of the aftereffect is weaker than the classical effect, and the aftereffect shows complete interocular transfer. Like the classic effect, the anti-McCollough effect (AME) is long lasting. Despite producing a less saturated illusory colour, the induction of an AME may override a previously induced ME.

Given that AMEs do transfer interocularly it is reasonable to suppose that they must occur in higher, binocular regions of the brain. Despite producing a less saturated illusory colour, the induction of an AME may override a previously induced ME, providing additional weight to the argument that AMEs occur in the higher visual areas than MEs Explanations of the effect by adaptation of edge-detectors, functional ECDs, and classical conditioning are compelling but will have to be adjusted for the inclusion of AMEs.

Missing Square Puzzle

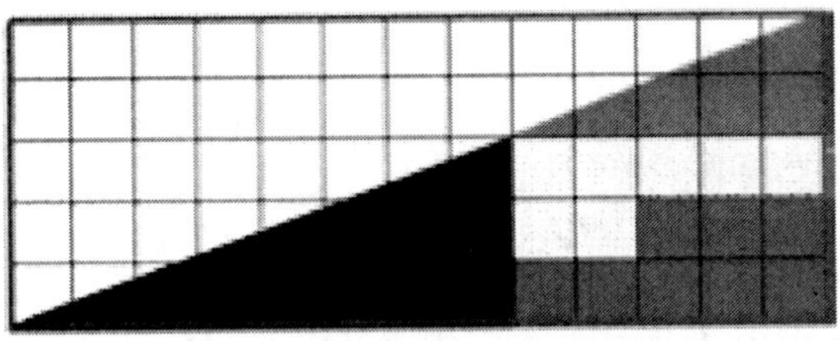

Figure: The missing square puzzle is an optical illusion used in mathematics classes to help students reason about geometrical figures. It depicts two arrangements of shapes, each of which apparently forms a 13×5 right-angled triangle, but one of which has a 1×1 hole in it.

Solution

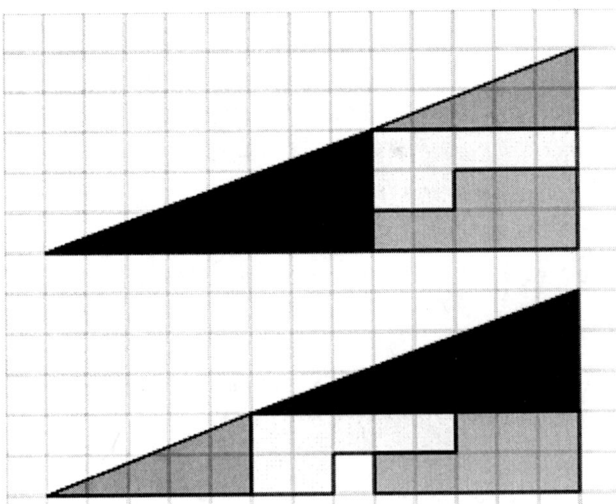

Figure: The key to the puzzle is the fact that neither of the 13×5 "triangles" is truly a triangle, because what would be the hypotenuse is bent. In other words, the hypotenuse does not maintain a consistent slope, even though it may appear that way to the human eye. A true 13 × 5 triangle cannot be created from the given component parts.

The four figures (the yellow, red, blue and green shapes) total 32 units of area, but the triangles are 13 wide and 5 tall, so it seems, that the area should be $S = \frac{13 \cdot 5}{2} = 32.5$ units. But the blue triangle has a ratio of 5:2 (=2.5:1), while the red triangle has the ratio 8:3 (H-2.667:1), and these are not the same ratio. So the apparent combined hypotenuse in each figure is actually bent. The amount of bending is around 1/28th of a unit (1.245364267°), which is difficult to see on the diagram of this puzzle. Note the grid point where the red and blue hypotenuses meet, and compare it to the same point on the other figure; the edge is slightly over or under the mark.

Overlaying the hypotenuses from both figures results in a very thin parallelogram with the area of exactly one grid square, the same area "missing" from the second figure. According to Martin Gardner, the puzzle was invented by a New York City amateur magician Paul Curry in 1953. The principle of a dissection paradox has however been known since the 1860s. The integer dimensions of the parts of the puzzle (2, 3, 5, 8, 13) are successive Fibonacci numbers. Many other geometric dissection puzzles are based on a few simple properties of the famous Fibonacci sequence.

Similar Puzzles

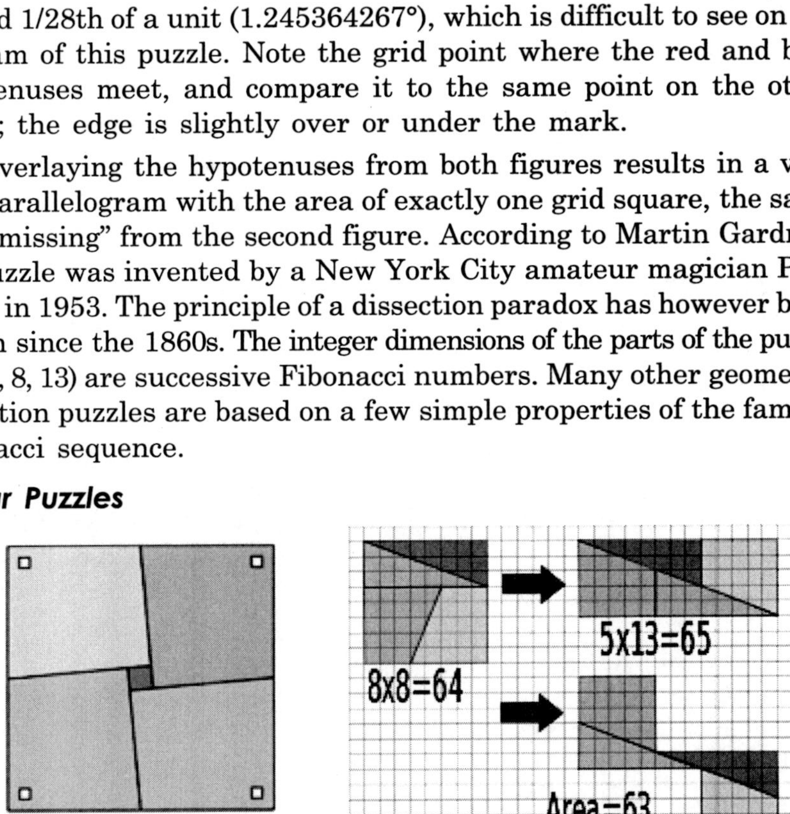

Figure: Sam Loyd's paradoxial dissection. In the "larger" rearrangement, the gaps between the figures have a combined unit square more area than their square gaps counterparts, creating an illusion that the figures there take up more space than those in the square figure. In the "smaller" rearrangement, the gaps take up one fewer unit squares than in the square.

A different puzzle of the same kind (depicted in the animation) uses four equal quadrilaterals and a small square, which form a larger square. When the quadrilaterals are rotated about their centres they fill the space of the small square, although the total area of the figure seems unchanged. The apparent paradox is explained by the fact that the side of the new large square is a little smaller than the original one. If a is the side of the large square and θ is the angle between two opposing sides in each quadrilateral, then the quotient between the two areas is given by sec ²θ " 1. For $\theta = 5°$, this is approximately 1.00765, which corresponds to a difference of about 0.8%.

Moon Illusion

The Moon illusion is an optical illusion in which the Moon appears larger near the horizon than it does while higher up in the sky. This optical illusion also occurs with the sun and star constellations. It has been known since ancient times, and recorded by numerous different cultures. The explanation of this illusion is still debated.

Proof of Illusion

A popular belief, stretching back at least to Aristotle in the 4th century B.C., holds that the Moon appears larger near the horizon due to a real magnification effect caused by the Earth's atmosphere. This is not true: although the atmosphere does change the perceived colour of the Moon, it does not magnify or enlarge it. In fact, the Moon appears about 1.5% smaller when it is near the horizon than when it is high in the sky, because it is farther away by up to one Earth radius and also because of atmospheric refraction, which makes the image of the Moon slightly smaller in the vertical axis. (Note that between different full moons, the Moon's angular diametre can vary from 33.5 arc minutes at perigee to 29.43 arc minutes at apogee—a difference of over 10%. This is because of the ellipticity of the Moon's orbit.) The angle that the full Moon subtends at an observer's eye can be measured directly with a theodoliteto show that it remains constant as the Moon rises or sinks in the sky (discounting the very small variations due to the physical effects mentioned). Photographs of the Moon at different elevations also show that its size remains the same.

A simple way of demonstrating that the effect is an illusion is to hold a small object (say, 1/4 inch wide) at arm's length (25 inches) with one eye closed, positioning it next to the seemingly large Moon. When the Moon is higher in the sky, positioning the same object near the Moon reveals that there is no change in size.

Possible Explanations

In the Book of Optics (1011-1022 C.E.), Ibn al-Haytham argued that vision occurs in the brain, and that personal experience has an effect on what people see and how they see, and that vision and perception are subjective. Arguing against Ptolemy's refraction theory, he redefined the problem in terms of perceived, rather than real, enlargement. He said that judging the distance of an object depends on there being an uninterrupted sequence of intervening bodies between the object and the observer. With the Moon, however, there are no intervening objects. Therefore, since the size of an object depends on its observed distance, which is in this case inaccurate, the Moon appears larger on the horizon. Through works by Roger Bacon, John Pecham and Witelo based on Ibn al-Haytham's explanation, the Moon illusion gradually came to be accepted as a psychological phenomenon, with Ptolemy's theory being rejected in the 17th century.

For over 100 years, research on the Moon illusion has been conducted by vision scientists who invariably have been psychologists specialising in human perception. After reviewing the many different explanations in their 2002 book The Mystery of the Moon Illusion, Ross and Plug conclude "No single theory has emerged victorious". The same conclusion is reached in the 1989 book, The Moon Illusion edited by Hershenson, which offers about 24 chapters written by different illusion researchers.

Angular Size and Physical Size

The "size" of an object in our view can be measured either as angular size (the angle that it subtends [is in opposition to] at the eye, corresponding to the proportion of the field of vision that it occupies) or physical size (its real size measured in, say, metres). As far as human perception is concerned, these two concepts are quite distinct. For example, if two small, identical, and familiar objects are placed at distances of five and ten metres respectively, then the more distant object subtends approximately half the angle of the nearer object, but we do not normally perceive that it is half the size. Conversely, if the more distant object did subtend the same angle as the nearer object then we would normally perceive it to be twice as big. A central question pertaining to the Moon illusion, therefore, is whether the horizon moon appears larger because its perceived angular size seems greater, or because its perceived physical size seems greater, or some combination of both. There is currently no firm consensus on this point.

Apparent Distance Hypothesis

An apparent distance theory evidently was first clearly described by Cleomedes around AD 200. The theory proposes that the horizon moon looks larger than the zenith moon because it looks farther away.

When we see objects such as clouds, birds and aeroplanes in the sky, those near the horizon are typically further away from us than those overhead. This may result in the perception of the sky itself as a comparatively flat or only gently curving surface in which objects moving towards the horizon always recede away from us.

If we perceive the Moon to be in the general vicinity of those other things we see in the sky, we would expect it to also recede as it approaches the horizon, which should result in a smaller retinal image. But since its retinal image is approximately the same size whether it is near the horizon or not, our brains, attempting to compensate for perspective, assume that a low moon must be physically larger.

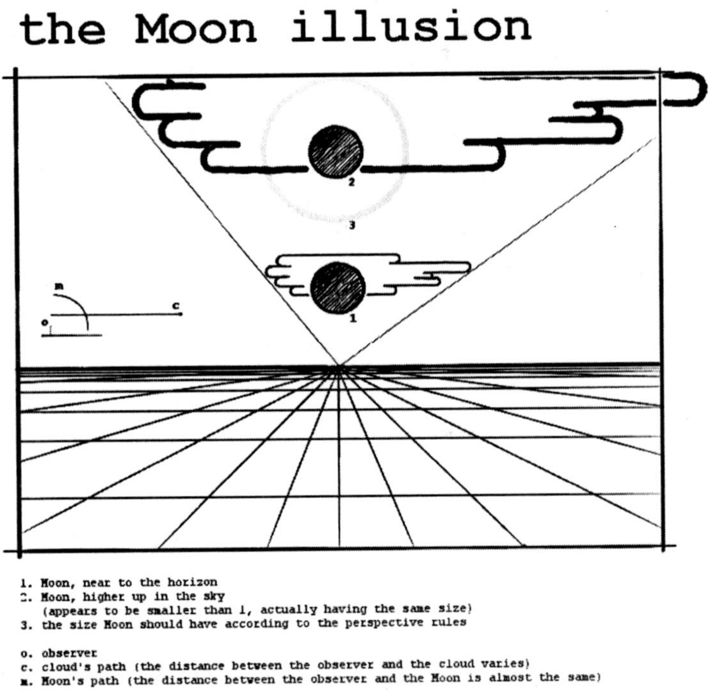

Figure: Comparison between the size of the Moon and the size of a cloud as they move higher in the sky.

Extensive experiments in 1962 by Kaufman and Rock showed that a crucial causative factor in the illusion is a change in the pattern of cues to distance. The horizon moon is perceived to be at the end of a stretch of terrain receding into the distance, accompanied by distant trees, buildings and so forth, all of which indicate that it must be a long way away, while these cues are absent from the zenith moon.

Experiments by many other researchers have found the same result; namely, when pictorial cues to a great distance are subtracted from the vista of the large-looking horizon moon it looks smaller. When pictorial cues to an increased distance are added into the vista of the zenith moon, it appears larger.

A potential problem for the apparent distance theory has been that very few people (perhaps about 5%) perceive the horizon moon as being both larger and farther away. Indeed most people (perhaps 90%) say the horizon moon looks both larger and *closer* than the zenith moon (Boring, 1962, Hershenson, 1982, McCready, 1965, 1986, Restle, 1970). Most of the rest say it looks larger and about the same distance away as the zenith moon, with a few people reporting no Moon illusion at all. Nevertheless, the apparent distance explanation is the one most often found in textbooks.

Those advocating the apparent distance hypothesis might counter that in the Ponzo illusion, people do not perceive one line to actually be farther away than another either, or that everything on the horizon is perceived as further away, out of the habit that this is true for every object ever seen there (average-altitude clouds are about 100 times further away at the horizon than when overhead).

Relative Size Hypothesis

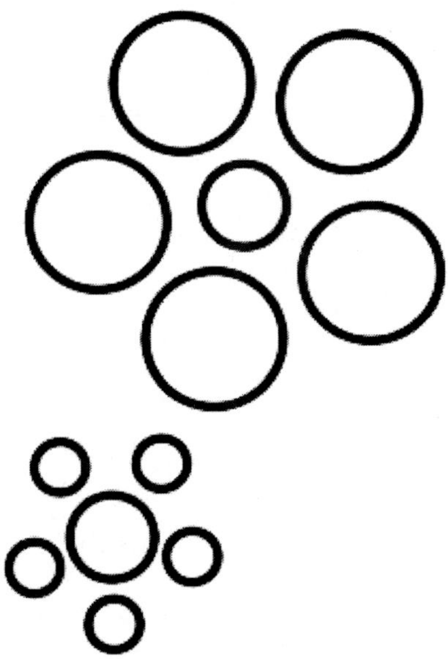

Figure: The Ebbinghaus illusion

Historically, the best-known alternative to the "apparent distance" theory has been a "relative size" theory. This states that the perceived size of an object depends not only on its retinal size, but also on the size of objects in its immediate visual environment. In the case of the Moon illusion, objects in the vicinity of the horizon moon (that is, objects on or near the horizon) exhibit a fine detail that makes the Moon appear larger, while the zenith moon is surrounded by large expanses of empty sky that make it appear smaller.

The effect is illustrated by the classic Ebbinghaus illusion shown at the right. The lower central circle surrounded by small circles might represent the horizon moon accompanied by objects of smaller visual extent, while the upper central circle represents the zenith moon surrounded by expanses of sky of larger visual extent. Although both central circles are actually the same size, to many people the lower one looks larger.

Angle of Regard Hypothesis

According to the angle of regard hypothesis, the Moon illusion is produced by changes in the position of the eyes in the head accompanying changes in the angle of elevation of the moon. Though once popular, this explanation no longer has much support.

Historical References to the Moon Illusion

Immanuel Kant refers to the moon illusion in his 1781 text the Critique of Pure Reason when he writes that "the astronomer cannot prevent himself from seeing the moon larger at its rising than some time afterwards, although he is not deceived by this illusion."

Motion Aftereffect

The motion after-effect (MAE) is a visual illusion experienced after viewing a moving visual stimulus for a time (seconds to minutes) with stationary eyes, and then fixating a stationary stimulus.

The stationary stimulus appears to move in the opposite direction to the original (physically moving) stimulus. The motion aftereffect is believed to be the result of motion adaptation.

For example, if one looks at a waterfall for about a minute and then looks at the stationary rocks at the side of the waterfall, these rocks appear to be moving upwards slightly. The illusory upwards movement is the motion aftereffect. This particular motion aftereffect is also known as the *waterfall illusion*. Another example can be seen when one looks at the centre of a rotating spiral for several seconds.

The spiral can exhibit outward or inward motion. When one then looks at any stationary pattern, it appears to be moving in the opposite direction. This form of the motion aftereffect is known as the spiral *aftereffect.*

Figure: Example movie which produces distortion illusion after one watches it and looks away.

Explanation

Neurons coding a particular movement reduce their responses with time of exposure to a constantly moving stimulus; this is neural adaptation. Neural adaptation also reduces the spontaneous, baseline activity of these same neurons when responding to a stationary stimulus. One theory is that perception of stationary objects, for example rocks beside a waterfall, is coded as the balance among the baseline responses of neurons coding all possible directions of motion. Neural adaptation of neurons stimulated by downwards movement reduces their baseline activity, tilting the balance in favour of upwards movement.

History

Aristotle (approx. 350 B.C.) reported illusory movement after viewing constant movement, but did not specify its direction. The first clear specification of the motion aftereffect was by Jan Evangelista Purkynμ (1820) who observed it after looking at a cavalry parade. Robert Addams (1834) reported the waterfall illusion after observing it at the Falls of Foyers in Scotland. According to Verstraten (1996) the term waterfall illusion was coined by Thompson (1888).

Mathematical Expression of the Principle

Consider the case of a point source located at a point P_0, vibrating at a frequency *f*. The disturbance may be described by a complex variable U_0 known as the complex amplitude.

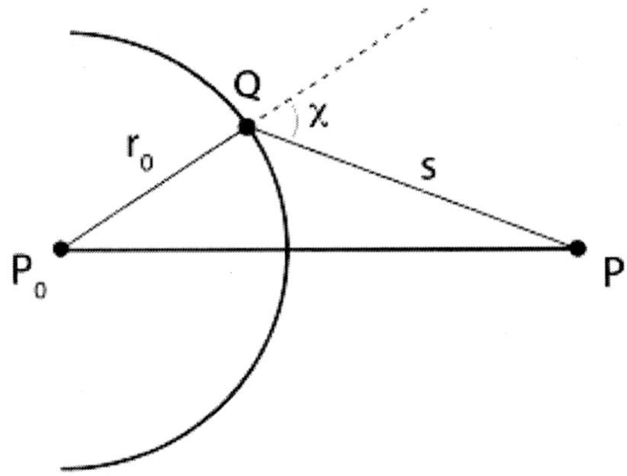

Figure: Geometric arrangement for Fresnel's calculation

It produces a spherical wave with wavelength λ, wavenumber $k = 2\pi/\lambda$. The complex amplitude of the primary wave at the point Q located at a distance r_0 from P_0 is given by

$$U(r_0) = \frac{U_0 e^{ikr_0}}{r_0}$$

since the magnitude decreases in inverse proportion to the distance travelled, and the phase changes as k times the distance travelled.

Using Huygens' theory and the principle of superposition of waves, the complex amplitude at a further point P is found by summing the contributions from each point on the sphere of radius r_0. In order to get agreement with experimental results, Fresnel found that the individual contributions from the secondary waves on the sphere had to be multiplied by a constant, i/λ, and by an additional inclination factor, $K(\chi)$. The first assumption means that the secondary waves oscillate at a quarter of a cycle out of phase with respect to the primary wave, and that the magnitude of the secondary waves are in a ratio of $1:\lambda$ to the primary wave. He also assumed that $K(\chi)$ had a maximum value when $\chi=0$, and was equal to zero when $\chi = \pi/2$. The complex amplitude at P is then given by:

$$U(P) = \frac{iU(r_0)}{\lambda} \int_S \frac{e^{iks}}{s} K(\chi) dS$$

where S describes the surface of the sphere, and s is the distance between Q and P. Fresnel used a zone construction method to find approximate values of K for the different zones, which enabled him to make predictions which were in agreement with experimental results. The various assumptions made by Fresnel emerge automatically in Kirchhoff's diffraction formula, to which the Huygens–Fresnel principle can be considered to be an approximation. Kirchoff gives the following expression for $K(\chi)$:

$$K(\chi) = -\frac{i}{2\lambda}(1 + \cos\chi)$$

This incorporates the quarter cycle phase shift and the reduced magnitude; K has a maximum value at $\chi = 0$ as in the Huygens–Fresnel principle; however, K is not equal to zero at $\chi = \pi/2$.

Huygens' Principle and Quantum Electrodynamics

Huygens' principle can be seen as a consequence of the isotropy of space - all directions in space are equal. Any disturbance created in a sufficiently small region of isotropic space (or in an isotropic medium) propagates from that region in all radial directions. The waves created by this disturbance, in turn, create disturbances in other regions, and so on. The superposition of all the waves results in the observed pattern of wave propagation. Isotropy of space is fundamental to quantum electrodynamics (QED) where the wave function of any object propagates along all available unobstructed paths. When integrated along all possible paths, with a phase factor which is proportional to the path length, the interference of the wave-functions correctly predicts observable phenomena.

Snell's Law

In optics and physics, Snell's law (also known as Descartes' law, the Snell–Descartes law, and the law of refraction) is a formula used to describe the relationship between the angles of incidence and refraction, when referring to light or other waves passing through a boundary between two different isotropic media, such as water and glass.

In optics, the law is used in ray tracing to compute the angles of incidence or refraction, and in experimental optics and gemology to find the refractive index of a material. The law is also satisfied in metamaterials, which allow light to be bent "backward" at a negative angle of refraction (negative refractive index). Although named after Dutch astronomer Willebrord Snellius(1580–1626), the law was first accurately described by the Persian scientist Ibn Sahl at Baghdad court, when in 984 he used the law to derive lens shapes that focus light with no geometric aberrations in the manuscript *On Burning Mirrors and Lenses* (984).

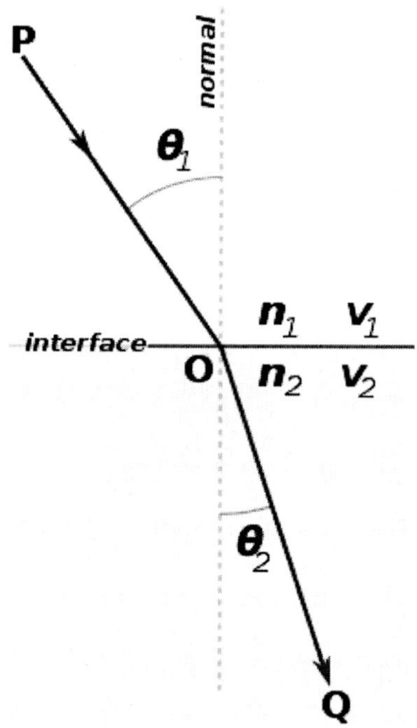

Figure : Refraction of light at the interface between two media of different refractive indices, with $n_2 > n_1$. Since the velocity is lower in the second medium ($v_2 < v_1$), the angle of refraction θ_2 is less than the angle of incidence θ_1; that is, the ray in the higher-index medium is closer to the normal.

Snell's law states that the ratio of the sines of the angles of incidence and refraction is equivalent to the ratio of phase velocities in the two media, or equivalent to the opposite ratio of the indices of refraction:

$$\frac{\sin\theta_1}{\sin\theta_2} = \frac{v_1}{v_2} = \frac{n_2}{n_1}$$

with each θ as the angle measured from the normal, v as the velocity of light in the respective medium (SI units are metres per second, or m/s) and n as the refractive index (which is unitless) of the respective medium.

The law follows from Fermat's principle of least time, which in turn follows from the propagation of light as waves.

History

Ptolemy, a Greek living in Alexandria, had found a relationship regarding refraction angles, but it was inaccurate for angles that were not small. Ptolemy was confident he had found an accurate empirical law, partially as a result of fudging his data to fit theory. Alhazen, in his *Book of Optics* (1021), came closer to discovering the law of refraction, though he did not take this step.

The law of refraction was first accurately described by Ibn Sahl, of Baghdad, in the manuscript *On Burning Mirrors and Lenses*(984). He made use of it to work out the shapes of lenses that focus light with no geometric aberrations, known as anaclastic lenses.

The law was rediscovered by Thomas Harriot in 1602, who however did not publish his results although he had corresponded with Kepler on this very subject. In 1621, Willebrord Snellius (Snel) derived a mathematically equivalent form, that remained unpublished during his lifetime.

René Descartes independently derived the law using heuristic momentum conservation arguments in terms of sines in his 1637 treatise *Discourse on Method*, and used it to solve a range of optical problems. Rejecting Descartes' solution, Pierre de Fermat arrived at the same solution based solely on his principle of least time.

According to Dijksterhuis, "In *De natura lucis et proprietate* (1662) Isaac Vossius said that Descartes had seen Snell's paper and concocted his own proof. We now know this charge to be undeserved but it has been adopted many times since." Both Fermat and Huygens repeated this accusation that Descartes had copied Snell. In French, Snell's Law is called "la loi de Descartes" or "loi de Snell-Descartes."

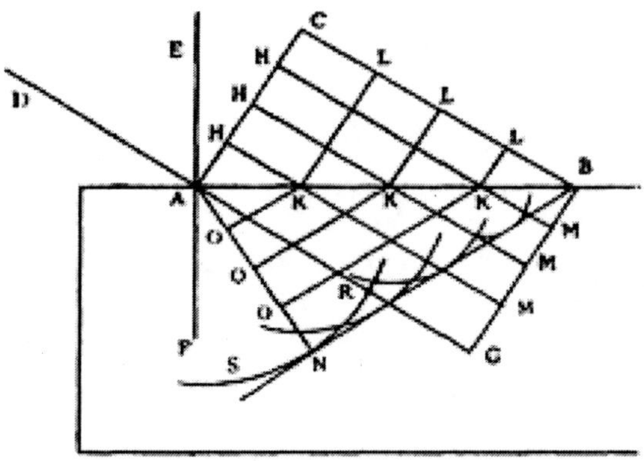

Figure : Christiaan Huygens' construction

In his 1678 *Traité de la Lumiere*, Christiaan Huygensshowed how Snell's law of sines could be explained by, or derived from, the wave nature of light, using what we have come to call the Huygens–Fresnel principle.

Explanation

Snell's law is used to determine the direction of light rays through refractive media with varying indices of refraction. The indices of refraction of the media, labelled n_1, n_2 and so on, are used to represent the factor by which a light ray's speed decreases when travelling through a refractive medium, such as glass or water, as opposed to its velocity in a vacuum. As light passes the border between media, depending upon the relative refractive indices of the two media, the light will either be refracted to a lesser angle, or a greater one. These angles are measured with respect to the *normal line*, represented perpendicular to the boundary. In the case of light travelling from air into water, light would be refracted towards the normal line, because the light is slowed down in water; light travelling from water to air would refract away from the normal line.

Refraction between two surfaces is also referred to as *reversible* because if all conditions were identical, the angles would be the same for light propagating in the opposite direction.

Snell's law is generally true only for isotropic or specular media (such as glass). In anisotropic media such as some crystals, birefringence may split the refracted ray into two rays, the *ordinary* or *o*-ray which follows Snell's law, and the other *extraordinary* or *e*-ray which may not be co-planar with the incident ray.

When the light or other wave involved is monochromatic, that is, of a single frequency, Snell's law can also be expressed in terms of a ratio of wavelengths in the two media, λ_1 and λ_2:

$$\frac{\sin\theta_1}{\sin\theta_2} = \frac{v_1}{v_2} = \frac{\lambda_1}{\lambda_2}$$

Derivations and Formulae

nell's law may be derived from Fermat's principle, which states that the light travels the path which takes the least time. By taking the derivative of the optical path length, the stationary point is found giving the path taken by the light (though it should be noted that the result does not show light taking the least time path, but rather one that is stationary with respect to small variations as there are cases where light actually takes the greatest time path, as in a spherical mirror). In a classic analogy, the area of lower refractive index is replaced by a beach, the area of higher refractive index by the sea, and the fastest way for a rescuer on the beach to get to a drowning person in the sea is to run along a path that follows Snell's law.

Alternatively, Snell's law can be derived using interference of all possible paths of light wave from source to observer—it results in destructive interference everywhere except extrema of phase (where interference is constructive)—which become actual paths.

Another way to derive Snell's Law involves an application of the general boundary conditions of Maxwell equations for electromagnetic radiation. Yet another way to derive Snell's law is based on translation symmetry considerations.

For example, a homogeneous surface perpendicular to the z direction can not change the transverse momentum. Since the propagation vector $\vec{k}$ is proportional to the photon's momentum, the transverse propagation direction $(k_x, k_y, 0)$ must remain the same in both regions. Assuming without loss of generality a plane of incidence in the z,x plane $k_{x\text{Region}_1} = k_{x\text{Region}_2}$. Using the well known dependence of the wave number on the refractive index of the medium, we derive Snell's law immediately.

$$k_{x\text{Region}_1} = k_{x\text{Region}_2}$$

$$n_1 k_0 \sin\theta_1 = n_2 k_0 \sin\theta_2$$

$$n_1 \sin\theta_1 = n_2 \sin\theta_2$$

where $k_0 = \frac{2\pi}{\lambda_0} = \frac{\omega}{c}$ is the wavenumber in vacuum. Note that no surface is truly homogeneous, in the least at the atomic scale. Yet full translational symmetry is an excellent approximation whenever the region is homogeneous on the scale of the light wavelength.

Vector Form

Given a normalised light vector l (pointing from the light source toward the surface) and a normalised plane normal vector n, one can work out the normalised reflected and refracted rays:

$$\cos\theta_1 = \mathbf{n}\cdot(-\mathbf{l})$$

$$\cos\theta_2 = \sqrt{1 - \left(\frac{n_1}{n_2}\right)^2 \left(1 - \left(\cos\theta_1\right)^2\right)}$$

$$\mathbf{v}_{\text{reflect}} = \mathbf{l} + \left(2\cos\theta_1\right)\mathbf{n}$$

$$\mathbf{v}_{\text{refract}} = \left(\frac{n_1}{n_2}\right)\mathbf{l} + \left(\frac{n_1}{n_2}\cos\theta_1 - \cos\theta_2\right)\mathbf{n}$$

Note: $\mathbf{n}\cdot(-\mathbf{l})$ must be positive. Otherwise, use

$$\mathbf{v}_{\text{refract}} = \left(\frac{n_1}{n_2}\right)\mathbf{l} + \left(-\frac{n_1}{n_2}\cos\theta_1 + \cos\theta_2\right)\mathbf{n}.$$

Example:

$$\mathbf{l} = \{0.707107, -0.707107\}, \mathbf{n} = \{0,1\}, \frac{n_1}{n_2} = 0.9$$

$$\cos\theta_1 = 0.707107,\ \cos\theta_2 = 0.771362$$

$$\mathbf{v}_{\text{reflect}} = \{0.707107, 0.707107\},$$

$$\mathbf{v}_{\text{refract}} = \{0.636396, -0.771362\}$$

The cosines may be recycled and used in the Fresnel equations for working out the intensity of the resulting rays.

Total internal reflection is indicated by a negative radic and in the equation for cos θ_2.

In this case, an evanescent wave is produced, which rapidly decays from the surface into the second medium. Conservation of energy is maintained by the circulation of energy across the boundary, averaging to zero net energy transmission.

Total Internal Reflection and Critical Angle

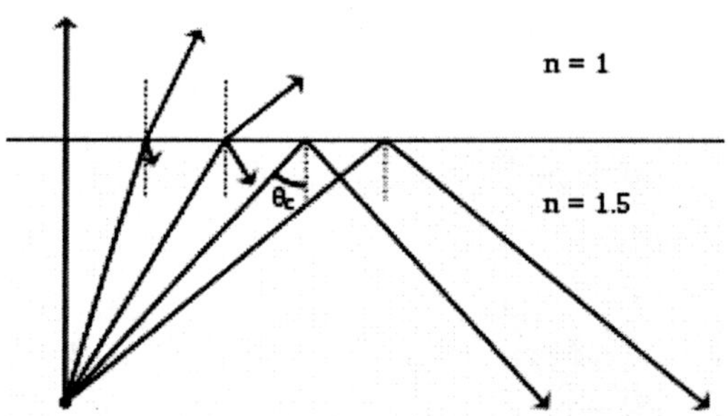

Figure: Demonstration of no refraction at angles greater than the critical angle.

When light travels from a medium with a higher refractive index to one with a lower refractive index, Snell's law seems to require in some cases (whenever the angle of incidence is large enough) that the sine of the angle of refraction be greater than one. This of course is impossible, and the light in such cases is completely reflected by the boundary, a phenomenon known as total internal reflection. The largest possible angle of incidence which still results in a refracted ray is called the critical angle; in this case the refracted ray travels along the boundary between the two media. For example, consider a ray of light moving from water to air with an angle of incidence of 50°. The refractive indices of water and air are approximately 1.333 and 1, respectively, so Snell's law gives us the relation

$$\sin\theta_2 = \frac{n_1}{n_2}\sin\theta_1 = 1.333 \cdot 0.766 = 1.021,$$

which is impossible to satisfy. The critical angle θ_{crit} is the value of θ_1 for which θ_2 equals 90°:

$$\theta_{crit} = \arcsin\left(\frac{n_2}{n_1}\sin\theta_2\right) = \arcsin\frac{n_2}{n_1} = 48.6^{\circ}.$$

Dispersion

In many wave-propagation media, wave velocity changes with frequency or wavelength of the waves; this is true of light propagation in most transparent substances other than a vacuum. These media are called dispersive. The result is that the angles determined by Snell's law also depend on frequency or wavelength, so that a ray of mixed wavelengths, such as white light, will spread or disperse. Such dispersion of light in glass or water underlies the origin of rainbows and other optical phenomena, in which different wavelengths appear as different colours. In optical instruments, dispersion leads to chromatic aberration; a colour-dependent blurring that sometimes is the resolution-limiting effect. This was especially true in refracting telescopes, before the invention of achromatic objective lenses.

Lossy, Absorbing, or Conducting Media

In a conducting medium, permittivity and index of refraction are complex-valued. Consequently, so are the angle of refraction and the wave-vector. This implies that, while the surfaces of constant real phase are planes whose normals make an angle equal to the angle of refraction with the interface normal, the surfaces of constant amplitude, in contrast, are planes parallel to the interface itself. Since these two planes do not in general coincide with each other, the wave is said to be inhomogeneous. The refracted wave is exponentially attenuated, with exponent proportional to the imaginary component of the index of refraction.

Specular Reflection

Specular reflection is the mirror-like reflection of light (or of other kinds of wave) from a surface, in which light from a single incoming direction (a ray) is reflected into a single outgoing direction. Such behaviour is described by the law of reflection, which states that the direction of incoming light (the incident ray), and the direction of outgoing light reflected (the reflected ray) make the same angle with respect to the surface normal, thus *the angle of incidence equals the angle of reflection* ($\theta_i = \theta_r$ in the figure), and that the incident, normal, and reflected directions are coplanar.

This behaviour was first discovered through careful observation and measurement by Hero of Alexandria (AD c. 10–70).

Explanation

Specular reflection is distinct from diffuse reflection, where incoming light is reflected in a broad range of directions. An example of the distinction between specular and diffuse reflection would be glossy and matte paints. Matte paints have almost exclusively diffuse reflection, while glossy paints have both specular and diffuse reflection. A surface built from a non-absorbing powder, such as plaster, can be a nearly perfect diffuser. On the opposite side, polished metallic objects can specularly reflect light very efficiently. The reflecting material of mirrors is usually aluminium or silver.

Even when a surface exhibits only specular reflection with no diffuse reflection, not all of the light is necessarily reflected. Some of the light may be absorbed by the materials. Additionally, depending on the type of material behind the surface, some of the light may be transmitted through the surface.

For most interfaces between materials, the fraction of the light that is reflected increases with increasing angle of incidence θ_i. If the light is propagating in a material with a higher index of refraction than the material whose surface it strikes, then total internal reflection may occur if the angle of incidence is greater than a certain critical angle. Specular reflection from a dielectric such as water can affect polarisation and at Brewster's angle reflected light is completely linearly polarised parallel to the interface.

The law of reflection arises from diffraction of a plane wave with small wavelength on a flat boundary: when the boundary size is much larger than the wavelength then electrons of the boundary are seen oscillating exactly in phase only from one direction – the specular direction. If a mirror becomes very small compared to the wavelength, the law of reflection no longer holds and the behaviour of light is more complicated.

Waves other than visible light can also exhibit specular reflection. This includes other electromagnetic waves, as well as non-electromagnetic waves. Examples include ionospheric reflection of radiowaves, reflection of radio- or microwave radar signals by flying objects, acoustic mirrors, which reflect sound, and atomic mirrors, which reflect neutral atoms. For the efficient reflection of atoms from a solid-statemirror, very cold atoms and/or grazing incidence are used in order to provide significant quantum reflection; ridged mirrors are used to enhance the specular reflection of atoms.

The reflectivity of a surface is the ratio of reflected power to incident power. The reflectivity is a material characteristic, depends on the wavelength, and is related to the refractive index of the material through Fresnel's equations.

In absorbing materials, like metals, it is related to the electronic absorption spectrum through the imaginary component of the complex refractive index. Measurements of specular reflection are performed with normal or varying incidence reflectometres using a scanning variable-wavelength light source. Lower quality measurements using a glossmetre quantify the glossy appearance of a surface in gloss units.

The image in a flat mirror has these features:

It is the same distance behind the mirror as the object is in front.

It is the same size as the object.

It is the right way up (erect).

It appears to be laterally inverted, in other words left and right reversed.

It is *virtual*, meaning that the image appears to be behind the mirror, and cannot be projected onto a screen.

Direction of Reflection

The direction of a reflected ray is determined by the vector of incidence and the surface normal vector. Given an incident direction $\hat{\mathbf{d}}_i$ from the surface to the light source and the surface normal direction $\hat{\mathbf{d}}_n$, the specularly reflected direction $\hat{\mathbf{d}}_s$ (all unit vectors) is:

$$\hat{\mathbf{d}}_s = 2\left(\hat{\mathbf{d}}_n \cdot \hat{\mathbf{d}}_i\right)\hat{\mathbf{d}}_n - \hat{\mathbf{d}}_i,$$

where $\hat{\mathbf{d}}_n \cdot \hat{\mathbf{d}}_i$ is a scalar obtained with the dot product. Different authors may define the incident and reflection directions with different signs. Assuming these Euclidean vectors are represented in column form, the equation can be equivalently expressed as a matrix-vector multiplication:

$$\hat{\mathbf{d}}_s = \mathbf{R}\,\hat{\mathbf{d}}_i,$$

where $\mathbf{R}$ is the so-called Householder transformation matrix, defined as:

$$\mathbf{R} = 2\hat{\mathbf{d}}_n\hat{\mathbf{d}}_n^{T} - \mathbf{I};$$

T denotes transposition and $\mathbf{I}$ is the identity matrix.

Hamilton's Principle

In physics, Hamilton's principle is William Rowan Hamilton's formulation of the principle of stationary action. It states that the dynamics of a physical system is determined by a variational problem for a functional based on a single function, the Lagrangian, which contains all physical information concerning the system and the forces acting on it. The variational problem is equivalent to and allows for the derivation of the *differential* equations of motion of the physical system. Although formulated originally for classical mechanics, Hamilton's principle also applies to classical fields such as the electromagnetic and gravitational fields, and has even been extended to quantum mechanics, quantum field theory and criticality theories.

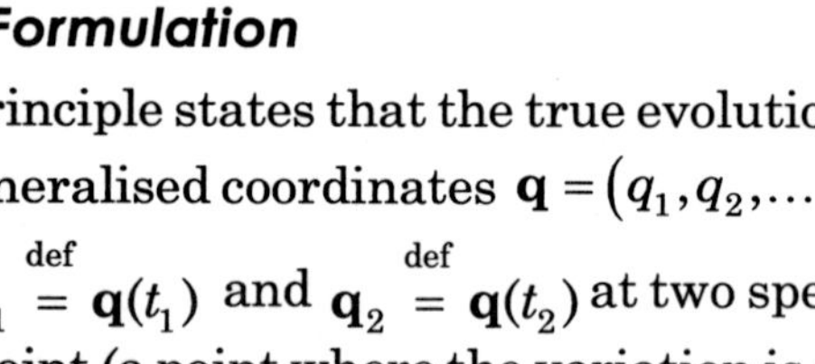

Mathematical Formulation

Hamilton's principle states that the true evolution $\mathbf{q}(t)$ of a system described by N generalised coordinates $\mathbf{q} = (q_1, q_2, \ldots, q_N)$ between two specified states $\mathbf{q}_1 \overset{\text{def}}{=} \mathbf{q}(t_1)$ and $\mathbf{q}_2 \overset{\text{def}}{=} \mathbf{q}(t_2)$ at two specified times t_1 and t_2 is a stationary point (a point where the variation is zero), of the action functional

$$\mathcal{S}[\mathbf{q}] \overset{\text{def}}{=} \int_{t_1}^{t_2} L(\mathbf{q}(t), \dot{\mathbf{q}}(t), t)\,dt$$

where $L(\mathbf{q}, \dot{\mathbf{q}}, t)$ is the Lagrangian function for the system. In other words, any *first-order perturbation* of the true evolution results in (at most) *second-order* changes in $\mathcal{S}$. The action $\mathcal{S}$ is a functional, i.e., something that takes as its input a function and returns a single number, a scalar.

In terms of functional analysis, Hamilton's principle states that the true evolution of a physical system is a solution of the functional equation

$$\frac{\delta \mathcal{S}}{\delta \mathbf{q}(t)} = 0$$

Euler-Lagrange Equations for the Action Integral

Requiring that the true trajectory $\mathbf{q}(t)$ be a stationary point of the action functional is $\mathcal{S}$ equivalent to a set of differential equations for $\mathbf{q}(t)$ (the Euler-Lagrange equations), which may be derived as follows. Let $\mathbf{q}(t)$ represent the true evolution of the system between two specified states $\mathbf{q}_1 \overset{\text{def}}{=} \mathbf{q}(t_1)$ and $\mathbf{q}_2 \overset{\text{def}}{=} \mathbf{q}(t_2)$ at two specified times t_1 and t_2, and let $\boldsymbol{\varepsilon}(t)$ be a small perturbation that is zero at the endpoints of the trajectory

$$\boldsymbol{\varepsilon}(t_1) = \boldsymbol{\varepsilon}(t_2) \overset{\text{def}}{=} 0$$

To first order in the perturbation $\boldsymbol{\varepsilon}(t)$, the change in the action functional $\delta\mathcal{S}$ would be

$$\delta \mathcal{S} = \int_{t_1}^{t_2} \left[L(\mathbf{q} + \boldsymbol{\varepsilon}, \dot{\mathbf{q}} + \dot{\boldsymbol{\varepsilon}}) - L(\mathbf{q}, \dot{\mathbf{q}}) \right] dt = \int_{t_1}^{t_2} \left(\boldsymbol{\varepsilon} \cdot \frac{\partial L}{\partial \mathbf{q}} + \dot{\boldsymbol{\varepsilon}} \cdot \frac{\partial L}{\partial \dot{\mathbf{q}}} \right) dt$$

where we have expanded the Lagrangian L to first order in the perturbation $\boldsymbol{\varepsilon}(t)$.

Applying integration by parts to the last term results in

$$\delta \mathcal{S} = \left[\boldsymbol{\varepsilon} \cdot \frac{\partial L}{\partial \dot{\mathbf{q}}} \right]_{t_1}^{t_2} + \int_{t_1}^{t_2} \left(\boldsymbol{\varepsilon} \cdot \frac{\partial L}{\partial \mathbf{q}} - \boldsymbol{\varepsilon} \cdot \frac{d}{dt} \frac{\partial L}{\partial \dot{\mathbf{q}}} \right) dt$$

The boundary conditions $\boldsymbol{\varepsilon}(t_1) = \boldsymbol{\varepsilon}(t_2) \overset{\text{def}}{=} 0$ causes the first term to vanish

$$\delta \mathcal{S} = \int_{t_1}^{t_2} \boldsymbol{\varepsilon} \cdot \left(\frac{\partial L}{\partial \mathbf{q}} - \frac{d}{dt} \frac{\partial L}{\partial \dot{\mathbf{q}}} \right) dt$$

Hamilton's principle requires that this first-order change $\delta\mathcal{S}$ is zero for all possible perturbations $\boldsymbol{\varepsilon}(t)$, i.e., the true path is a stationary point of the action functional $\mathcal{S}$ (either a minimum, maximum or saddle point). This requirement can be satisfied if and only if $\frac{\partial L}{\partial \mathbf{q}} - \frac{d}{dt}\frac{\partial L}{\partial \dot{\mathbf{q}}} = 0$

Euler-Lagrange equations

These equations are called the Euler-Lagrange equations for the variational problem.

The conjugate momentum p_k for a generalised coordinate q_k is defined by the equation $p_k \stackrel{\text{def}}{=} \frac{\partial L}{\partial \dot{q}_k}$.

An important special case of these equations occurs when L does not contain a generalised coordinate q_k explicitly, i.e.,

if $\frac{\partial L}{\partial q_k} = 0$, the conjugate momentum $p_k \stackrel{\text{def}}{=} \frac{\partial L}{\partial \dot{q}_k}$ is constant.

In such cases, the coordinate q_k is called a cyclic coordinate. For example, if we use polar coordinates t, r, θ to describe the planar motion of a particle, and if L does not depend on θ, the conjugate momentum is the conserved angular momentum.

Example: Free Particle in Polar Coordinates

Trivial examples help to appreciate the use of the action principle via the Euler-Lagrangian equations. A free particle (mass m and velocity v) in Euclidean space moves in a straight line. Using the Euler-Lagrange equations, this can be shown in polar coordinates as follows. In the absence of a potential, the Lagrangian is simply equal to the kinetic energy

$$L = \frac{1}{2}mv^2 = \frac{1}{2}m\left(\dot{x}^2 + \dot{y}^2\right)$$

in orthonormal (x,y) coordinates, where the dot represents differentiation with respect to the curve parameter (usually the time, t).

Therefore, upon application of the Euler-Lagrange equations,

$$\frac{d}{dt}\left(\frac{\partial L}{\partial \dot{x}}\right) - \frac{\partial L}{\partial x} = 0 \qquad \Rightarrow \qquad m\ddot{x} = 0$$

And likewise for y. Thus the Euler-Lagrange formulation can be used to derive Newton's laws.

In polar coordinates (r, φ) the kinetic energy and hence the Lagrangian becomes

$$L = \frac{1}{2}m\left(\dot{r}^2 + r^2\dot{\varphi}^2\right).$$

The radial r and φ components of the Euler-Lagrangian equations become, respectively

$$\frac{d}{dt}\left(\frac{\partial L}{\partial \dot{r}}\right) - \frac{\partial L}{\partial r} = 0 \quad \Rightarrow \quad \ddot{r} - r\dot{\varphi}^2 = 0$$

$$\frac{d}{dt}\left(\frac{\partial L}{\partial \dot{\varphi}}\right) - \frac{\partial L}{\partial \varphi} = 0 \quad \Rightarrow \quad \ddot{\varphi} + \frac{2}{r}\dot{r}\dot{\varphi} = 0.$$

The solution of these two equations is given by

$$r = \sqrt{(at+b)^2 + c^2}$$

$$\varphi = \arctan\left(\frac{at+b}{c}\right) + d$$

for a set of constants *a, b, c, d* determined by initial conditions. Thus, indeed, *the solution is a straight line* given in polar coordinates: *a* is the velocity, *c* is the distance of the closest approach to the origin, and *d* is the angle of motion.

Comparison with Maupertuis' Principle

Hamilton's principle and Maupertuis' principle are occasionally confused and both have been called (incorrectly) the principle of least action. They differ in three important ways:

their definition of the action...

Maupertuis' principle uses an integral over the generalised coordinates known as the abbreviated action $\mathcal{S}_0 \stackrel{\text{def}}{=} \int \mathbf{p} \cdot d\mathbf{q}$ where $\mathbf{p} = (p_1, p_2, \ldots, p_N)$ are the conjugate momenta defined above. By contrast, Hamilton's principle uses , the integral of the Lagrangian over time.

the solution that they determine...

Hamilton's principle determines the trajectory $\mathbf{q}(t)$ as a function of time, whereas Maupertuis' principle determines only the shape of the trajectory in the generalised coordinates. For example, Maupertuis' principle determines the shape of the ellipse on which a particle moves under the influence of an inverse-square central force such as gravity, but does not describe *per se* how the particle moves along that trajectory. (However, this time parameterization may be determined from the trajectory itself in subsequent calculations using the conservation of energy.) By contrast, Hamilton's principle directly specifies the motion along the ellipse as a function of time.

...and the constraints on the variation.

Maupertuis' principle requires that the two endpoint states q_1 and q_2 be given and that energy be conserved along every trajectory. By contrast, Hamilton's principle does not require the conservation of energy, but does require that the endpoint times t_1 and t_2 be specified as well as the endpoint states q_1 and q_2.

Action Principle for Classical Fields

The action principle can be extended to obtain the equations of motion for fields, such as the electromagnetic field or gravity. The Einstein equation utilises the *Einstein-Hilbert action* as constrained by a variational principle. The path of a body in a gravitational field (i.e. free fall in space time, a so called geodesic) can be found using the action principle.

Hamilton's Principle Applied to Deformable Bodies

Hamilton's principle is an important variational principle in elastodynamics. As opposed to a system composed of rigid bodies, deformable bodies have an infinite number of degrees of freedom and occupy continuous regions of space; consequently, the state of the system is described by using continuous functions of space and time. The extended Hamilton Principle for such bodies is given by

$$\int_{t1}^{t2}\left[\delta W_e + \delta T - \delta U\right]dt = 0$$

where T is the kinetic energy, U is the elastic energy, W_e is the work done by external loads on the body, and t_1, t_2 the initial and final times. If the system is conservative, the work done by external forces may be derived from a scalar potential V. In this case,

$$\delta\int_{t1}^{t2}\left[T - (U + V)\right]dt = 0.$$

This is called Hamilton's principle and it is invariant under coordinate transformations.

Action Principle in Quantum Mechanics and Quantum Field Theory

In quantum mechanics, the system does not follow a single path whose action is stationary, but the behaviour of the system depends on all imaginable paths and the value of their action. The action corresponding to the various paths is used to calculate the path integral, that gives the probability amplitudes of the various outcomes.

Although equivalent in classical mechanics with Newton's laws, the action principle is better suited for generalisations and plays an important role in modern physics. Indeed, this principle is one of the great generalisations in physical science. In particular, it is fully appreciated and best understood within quantum mechanics. Richard Feynman's path integral formulation of quantum mechanics is based on a stationary-action principle, using path integrals. Maxwell's equations can be derived as conditions of stationary action.

Classical Mechanics

In physics, classical mechanics is one of the two major sub-fields of mechanics, which is concerned with the set of physical laws describing the motion of bodies under the action of a system of forces. The study of the motion of bodies is an ancient one, making classical mechanics one of the oldest and largest subjects in science, engineering and technology.

Classical mechanics describes the motion of macroscopic objects, from projectiles to parts of machinery, as well as astronomical objects, such as spacecraft, planets, stars, and galaxies. Besides this, many specialisations within the subject deal with gases, liquids, solids, and other specific sub-topics. Classical mechanics provides extremely accurate results as long as the domain of study is restricted to large objects and the speeds involved do not approach the speed of light.

When the objects being dealt with become sufficiently small, it becomes necessary to introduce the other major sub-field of mechanics, quantum mechanics, which reconciles the macroscopic laws of physics with the atomic nature of matter and handles the wave-particle duality of atoms and molecules. In the case of high velocity objects approaching the speed of light, classical mechanics is enhanced by special relativity. General relativity unifies special relativity with Newton's law of universal gravitation, allowing physicists to handle gravitation at a deeper level.

The term *classical mechanics* was coined in the early 20th century to describe the system of physics begun by Isaac Newton and many contemporary 17th century natural philosophers, building upon the earlier astronomical theories of Johannes Kepler, which in turn were based on the precise observations of Tycho Brahe and the studies of terrestrial projectile motion of Galileo. Because these aspects of physics were developed long before the emergence of quantum physics and relativity, some sources exclude Einstein's theory of relativity from this category. However, a number of modern sources *do* include relativistic mechanics, which in their view represents *classical mechanics* in its most developed and most accurate form.

The initial stage in the development of classical mechanics is often referred to as Newtonian mechanics, and is associated with the physical concepts employed by and the mathematical methods invented by Newton himself, in parallel with Leibniz, and others.

This is further described in the following sections. Later, more abstract and general methods were developed, leading to reformulations of classical mechanics known as Lagrangian mechanics and Hamiltonian mechanics.

These advances were largely made in the 18th and 19th centuries, and they extend substantially beyond Newton's work, particularly through their use of analytical mechanics. Ultimately, the mathematics developed for these were central to the creation of quantum mechanics.

Description of the theory

The following introduces the basic concepts of classical mechanics. For simplicity, it often models real-world objects as point particles, objects with negligible size. The motion of a point particle is characterised by a small number of parameters: its position, mass, and the forces applied to it. Each of these parameters is discussed in turn.

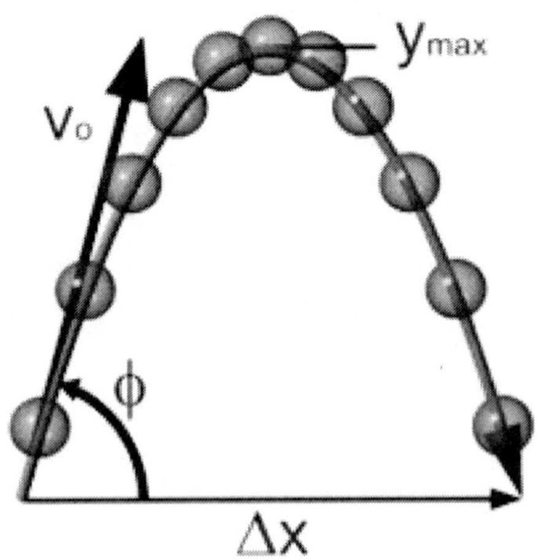

Figure: The analysis of projectile motion is a part of classical mechanics.

In reality, the kind of objects that classical mechanics can describe always have a non-zero size. (The physics of *very* small particles, such as the electron, is more accurately described by quantum mechanics). Objects with non-zero size have more complicated behaviour than hypothetical point particles, because of the additional degrees of freedom—for example, a baseball can spin while it is moving.

However, the results for point particles can be used to study such objects by treating them as composite objects, made up of a large number of interacting point particles. The centre of mass of a composite object behaves like a point particle.

Position and its Derivatives

The SI derived "mechanical" (that is, not electromagnetic or thermal) units with kg, m and s

Position	M
Angular position/Angle	unitless (radian)
velocity	$m\ s^{-1}$
Angular velocity	s^{-1}
acceleration	$m\ s^{-2}$
Angular acceleration	s^{-2}
jerk	$m\ s^{-3}$
"Angular jerk"	s^{-3}
specific energy	$m^2\ s^{-2}$
absorbed dose rate	$m^2\ s^{-3}$
moment of inertia	$kg\ m^2$
momentum	$kg\ m\ s^{-1}$
angular momentum	$kg\ m^2\ s^{-1}$
force	$kg\ m\ s^{-2}$
torque	$kg\ m^2\ s^{-2}$
energy	$kg\ m^2\ s^{-2}$
power	$kg\ m^2\ s^{-3}$
pressure and energy density	$kg\ m^{-1}\ s^{-2}$
surface tension	$kg\ s^{-2}$
Spring constant	$kg\ s^{-2}$
irradiance and energy flux	$kg\ s^{-3}$
kinematic viscosity	$m^2\ s^{-1}$
dynamic viscosity	$kg\ m^{-1}\ s^{-1}$

Density (mass density)	kg m^{-3}
Density (weight density)	kg m^{-2} s^{-2}
Number density	m^{-3}
Action	kg m^2 s^{-1}

The *position* of a point particle is defined with respect to an arbitrary fixed reference point, O, in space, usually accompanied by a coordinate system, with the reference point located at the *origin* of the coordinate system. It is defined as the vector r from O to the particle. In general, the point particle need not be stationary relative to O, so r is a function of *t*, the time elapsed since an arbitrary initial time. In pre-Einstein relativity (known as Galilean relativity), time is considered an absolute, i.e., the time interval between any given pair of events is the same for all observers. In addition to relying on absolute time, classical mechanics assumes Euclidean geometry for the structure of space.

Velocity and Speed

The *velocity*, or the rate of change of position with time, is defined as the derivative of the position with respect to time or

$$\mathbf{v} = \frac{d\mathbf{r}}{dt} .$$

In classical mechanics, velocities are directly additive and subtractive. For example, if one car travelling East at 60 km/h passes another car travelling East at 50 km/h, then from the perspective of the slower car, the faster car is travelling east at 60 " 50 = 10 km/h. Whereas, from the perspective of the faster car, the slower car is moving 10 km/h to the West. Velocities are directly additive as vector quantities; they must be dealt with using vector analysis. Mathematically, if the velocity of the first object in the previous discussion is denoted by the vector u = *u*d and the velocity of the second object by the vector v = *v*e, where *u* is the speed of the first object, *v* is the speed of the second object, and d and e are unit vectors in the directions of motion of each particle respectively, then the velocity of the first object as seen by the second object is

$$\mathbf{u}' = \mathbf{u} - \mathbf{v}.$$

Similarly,

$$\mathbf{v}' = \mathbf{v} - \mathbf{u}.$$

When both objects are moving in the same direction, this equation can be simplified to

$$\mathbf{u}' = (u - v)\mathbf{d}.$$

Or, by ignoring direction, the difference can be given in terms of speed only:

$$u' = u - v.$$

Acceleration

The *acceleration*, or rate of change of velocity, is the derivative of the velocity with respect to time (the second derivative of the position with respect to time) or

$$\mathbf{a} = \frac{d\mathbf{v}}{dt}.$$

Acceleration can arise from a change with time of the magnitude of the velocity or of the direction of the velocity or both. If only the magnitude v of the velocity decreases, this is sometimes referred to as deceleration, but generally any change in the velocity with time, including deceleration, is simply referred to as acceleration.

Frames of Reference

While the position and velocity and acceleration of a particle can be referred to any observer in any state of motion, classical mechanics assumes the existence of a special family of reference frames in terms of which the mechanical laws of nature take a comparatively simple form. These special reference frames are called inertial frames. An inertial frame is such that when an object without any force interactions (an idealised situation) is viewed from it, it will appear either to be at rest or in a state of uniform motion in a straight line. This is the fundamental definition of an inertial frame.

They are characterised by the requirement that all forces entering the observer's physical laws originate in identifiable sources (charges, gravitational bodies, and so forth). A non-inertial reference frame is one accelerating with respect to an inertial one, and in such a non-inertial frame a particle is subject to acceleration by fictitious forces that enter the equations of motion solely as a result of its accelerated motion, and do not originate in identifiable sources.

These fictitious forces are in addition to the real forces recognised in an inertial frame. A key concept of inertial frames is the method for identifying them. For practical purposes, reference frames that are unaccelerated with respect to the distant starsare regarded as good approximations to inertial frames.

Consider two reference frames *S* and *S'* . For observers in each of the reference frames an event has space-time coordinates of (x,y,z,t) in frame *S* and $(x2,y2,z2,t2)$ in frame *S2*. Assuming time is measured the same in all reference frames, and if we require $x = x'$ when $t = 0$, then the relation between the space-time coordinates of the same event observed from the reference frames *S2* and *S*, which are moving at a relative velocity of u in the x direction is:

$$x2 = x - ut$$

$$y2 = y$$

$$z2 = z$$

$$t2 = t$$

This set of formulas defines a group transformation known as the Galilean transformation (informally, the *Galilean transform*). This group is a limiting case of the Poincaré group used in special relativity. The limiting case applies when the velocity u is very small compared to c, the speed of light.

The transformations have the following consequences:

- v2 = v - u (the velocity v2 of a particle from the perspective of *S*2 is slower by u than its velocity v from the perspective of *S*)
- a2 = a (the acceleration of a particle is the same in any inertial reference frame)
- F2 = F (the force on a particle is the same in any inertial reference frame)
- the speed of light is not a constant in classical mechanics, nor does the special position given to the speed of light in relativistic mechanics have a counterpart in classical mechanics.

For some problems, it is convenient to use rotating coordinates (reference frames). Thereby one can either keep a mapping to a convenient inertial frame, or introduce additionally a fictitious centrifugal force and Coriolis force.

Forces; Newton's Second Law

Newton was the first to mathematically express the relationship between force and momentum. Some physicists interpret Newton's second law of motion as a definition of force and mass, while others consider it to be a fundamental postulate, a law of nature. Either interpretation has the same mathematical consequences, historically known as "Newton's Second Law":

$$\mathbf{F} = \frac{\mathrm{d}\mathbf{p}}{\mathrm{d}t} = \frac{\mathrm{d}(m\mathbf{v})}{\mathrm{d}t}.$$

The quantity mv is called the (canonical) momentum. The net force on a particle is thus equal to rate change of momentum of the particle with time. Since the definition of acceleration is a = dv/dt, the second law can be written in the simplified and more familiar form:

$$\mathbf{F} = m\mathbf{a}.$$

So long as the force acting on a particle is known, Newton's second law is sufficient to describe the motion of a particle. Once independent relations for each force acting on a particle are available, they can be substituted into Newton's second law to obtain an ordinary differential equation, which is called the *equation of motion.*

As an example, assume that friction is the only force acting on the particle, and that it may be modelled as a function of the velocity of the particle, for example:

$$\mathbf{F}_{\mathrm{R}} = -\lambda\mathbf{v},$$

where λ is a positive constant. Then the equation of motion is

$$-\lambda\mathbf{v} = m\mathbf{a} = m\frac{\mathrm{d}\mathbf{v}}{\mathrm{d}t}.$$

This can be integrated to obtain

$$\mathbf{v} = \mathbf{v}_0 e^{-\lambda t/m}$$

where v_0 is the initial velocity. This means that the velocity of this particle decays exponentially to zero as time progresses. In this case, an equivalent viewpoint is that the kinetic energy of the particle is absorbed by friction (which converts it to heat energy in accordance with the conservation of energy), slowing it down. This expression can be further integrated to obtain the position r of the particle as a function of time.

Important forces include the gravitational force and the Lorentz force for electromagnetism. In addition, Newton's third law can sometimes be used to deduce the forces acting on a particle: if it is known that particle A exerts a force F on another particle B, it follows that B must exert an equal and opposite reaction *force*, "F, on A. The strong form of Newton's third law requires that F and "F act along the line connecting A and B, while the weak form does not. Illustrations of the weak form of Newton's third law are often found for magnetic forces.

Variational Principle

A variational principle is a scientific principle used within the calculus of variations, which develops general methods for finding functions which minimise or maximise the value of quantities that depend upon those functions. For example, to answer this question: "What is the shape of a chain suspended at both ends?" we can use the variational principle that the shape must minimise the gravitational potential energy.

According to Cornelius Lanczos, any physical law which can be expressed as a variational principle describes an expression which is self-adjoint. These expressions are also called Hermitian. Such an expression describes an invariant under a Hermitian transformation.

Felix Klein's Erlangen program attempted to identify such invariants under a group of transformations. In what is referred to in physics as Noether's theorem, the Poincaré group of transformations (what is now called a gauge group) for general relativity defines symmetries under a group of transformations which depend on a variational principle, or action principle.

Examples

- Variational principle of Ivar Ekeland
- Fermat's principle in geometrical optics.
- The principle of least action in mechanics, electromagnetic theory, and quantum mechanics.
- Maupertuis principle in classical mechanics.
- The Einstein equation also involves a variational principle, the Einstein-Hilbert action.
- Gauss' principle of least constraint.
- Hertz's principle of least curvature.

Variational Principle in Quantum Mechanics

In a system in which the Hamiltonian H, is known, if the Schrφdinger equation cannot be solved explicitly for the ground state wavefunction, then the ground state energy is bounded by the following: For any normalised wavefunction φ, the expectation value of the Hamiltonian for φ must be greater than or equal to the actual ground state energy. Or in other words:

$$E_{ground} \leq \langle \phi \mid H \mid \phi \rangle.$$

This holds for any trial φ since, by definition, the ground state wavefunction has the lowest energy, and any trial wavefunction will have energy greater than or equal to it.

Proof: φ can be expanded as a linear combination of the actual eigenfunctions of the Hamiltonian (which we assume to be normalised and orthogonal):

$$\phi = \sum_n c_n \psi_n.$$

Then, to find the expectation value of the hamiltonian:

$$\begin{aligned}\langle \phi | H | \phi \rangle &= \left\langle \sum_n c_n \psi_n | H | \sum_m c_m \psi_m \right\rangle \\ &= \sum_n \sum_m \langle c_n \psi_n | E_m | c_m \psi_m \rangle \\ &= \sum_n \sum_m c_n^* c_m E_m \langle \psi_n | \psi_m \rangle \\ &= \sum_n |c_n|^2 E_n.\end{aligned}$$

Now, the ground state energy is the lowest energy possible, i.e. $E_n \geq E_g$. Therefore, if the guessed wave function φ is normalised:

$$\langle \phi | H | \phi \rangle \geq E_g \sum_n |c_n|^2 = E_g.$$

In General

For a hamiltonian H that describes the studied system and *any* normalisable function ψ with arguments appropriate for the unknown wave function of the system, we define the functional

$$\varepsilon[\Psi] = \frac{\langle \Psi | \hat{H} | \Psi \rangle}{\langle \Psi | \Psi \rangle}.$$

The variational principle states that

- $\varepsilon \geq E_0$, where E_0 is the lowest energy eigenstate (ground state) of the hamiltonian
- $\varepsilon = E_0$ if and only if ψ is exactly equal to the wave function of the ground state of the studied system.

The variational principle formulated above is the basis of the variational method used in quantum mechanics and quantum chemistry to find approximations to the ground state.

The Harris functional method is anti-variational (it is a lower bound on the energy). Another facet in variational principles in quantum mechanics is that since ψ and $\psi^\dagger$ can be varied separately (a fact arising due to the complex nature of the wave function), the quantities can be varied in principle just one at a time.

Principle of Least Action

In physics, the principle of least action – or, more accurately, the principle of stationary action – is a variational principle that, when applied to the action of a mechanical system, can be used to obtain the equations of motion for that system. The principle led to the development of the Lagrangian and Hamiltonian formulations of classical mechanics.

The principle remains central in modern physics and mathematics, being applied in the theory of relativity, quantum mechanics and quantum field theory, and a focus of modern mathematical investigation in Morse theory. This article deals primarily with the historical development of the idea; a treatment of the mathematical description and derivation can be found in the article on the action. The chief examples of the principle of stationary action are Maupertuis' principle and Hamilton's principle.

The action principle is preceded by earlier ideas in surveying and optics. The rope stretchers of ancient Egypt stretched corded ropes between two points to measure the path which minimised the distance of separation, and Claudius Ptolemy, in his Geographia (Bk 1, Ch 2), emphasized that one must correct for "deviations from a straight course"; in ancient Greece Euclid states in his *Catoptrica* that, for the path of light reflecting from a mirror, the angle of incidence equals the angle of reflection; and Hero of Alexandria later showed that this path was the shortest length and least time. But the credit for the formulation of the principle as it applies to the action is often given to Pierre-Louis Moreau de Maupertuis, who wrote about it in 1744 and 1746. However, scholarship indicates that this claim of priority is not so clear; Leonhard Euler discussed the principle in 1744, and there is evidence that Gottfried Leibniz preceded both by 39 years.

Origins, Statement, and the Controversy

In the 17th century Pierre de Fermat postulated that "light travels between two given points along the path of shortest time," *which is known as the principle of least time or Fermat's principle.*

Maupertuis' Formulation

Credit for the formulation of the principle of least action is commonly given to Pierre Louis Maupertuis, who wrote about it in 1744 and 1746. Maupertuis felt that "Nature is thrifty in all its actions", and applied the principle broadly: The laws of movement and of rest deduced from this principle being precisely the same as those observed in nature, we can admire the application of it to all phenomena. The movement of animals, the vegetative growth of plants... are only its consequences; and the spectacle of the universe becomes so much the grander, so much more beautiful, the worthier of its Author, when one knows that a small number of laws, most wisely established, suffice for all movements.

This notion of Maupertuis, although somewhat deterministic today, does capture much of the essence of mechanics.

In application to physics, Maupertuis suggested that the quantity to be minimised was the product of the duration (time) of movement within a system by the "vis viva", twice what we now call the kinetic energy T of the system.

$$\int 2T(t) \cdot dt$$

Euler's Formulation

Leonhard Euler gave a formulation of the action principle in 1744, in very recognisable terms, in the Additamentum *2* to his *Methodus Inveniendi Lineas Curvas Maximi Minive Proprietate Gaudentes. He* begins the second paragraph :

Let the mass of the projectile be M, and let its speed be v while being moved over an infinitesimal distance ds. The body will have a momentum Mv that, when multiplied by the distance ds, will give $Mv\ ds$, the momentum of the body integrated over the distance ds. Now I assert that the curve thus described by the body to be the curve (from among all other curves connecting the same endpoints) that minimises

$$\int Mvds$$

or, provided that M is constant,

$$\int vds.$$

As Euler states,

$$\int Mvds$$

Is the integral of the momentum over distance travelled, which, in modern notation, equals the reduced action

$$\int pdq$$

Thus, Euler made an equivalent and (apparently) independent statement of the variational principle in the same year as Maupertuis, albeit slightly later. Curiously, Euler did not claim any priority, as the following episode shows.

Disputed Priority

Maupertuis' priority was disputed in 1751 by the mathematician Samuel Kφnig, who claimed that it had been invented by Gottfried Leibniz in 1707. Although similar to many of Leibniz's arguments, the principle itself has not been documented in Leibniz's works. Kφnig himself showed a *copy* of a 1707 letter from Leibniz to Jacob Hermann with the principle, but the *original* letter has been lost. In contentious proceedings, Kφnig was accused of forgery, and even the King of Prussia entered the debate, defending Maupertuis, while Voltaire defended Kφnig.

Euler, rather than claiming priority, was a staunch defender of Maupertuis, and Euler himself prosecuted Kφnig for forgery before the Berlin Academy on 13 April 1752. The claims of forgery were re-examined 150 years later, and archival work by C.I. Gerhardt in 1898 and W. Kabitz in 1913 uncovered other copies of the letter, and three others cited by Kφnig, in the Bernoulli archives.

Further Development

Euler continued to write on the topic; in his Reflexions sur quelques loix generales de la nature (1748), he called the quantity "effort". His expression corresponds to what we would now call potential energy, so that his statement of least action in statics is equivalent to the principle that a system of bodies at rest will adopt a configuration that minimises total potential energy.

Lagrange and Hamilton

Much of the calculus of variations was stated by Joseph Louis Lagrange in 1760 and he proceeded to apply this to problems in dynamics. In *Méchanique Analytique* (1788) Lagrange derived the general equations of motion of a mechanical body. William Rowan Hamilton in 1834 and 1835 applied the variational principle to the classical Lagrangian function $L = T - V$ to obtain the Euler-Lagrange equations in their present form.

Jacobi and Morse

In 1842, Carl Gustav Jacobi tackled the problem of whether the variational principle always found minima as opposed to other stationary points (maxima or stationary saddle points); most of his work focused on geodesics on two-dimensional surfaces. The first clear general statements were given by Marston Morse in the 1920s and 1930s, leading to what is now known as Morse theory. For example, Morse showed that the number of conjugate points in a trajectory equalled the number of negative eigenvalues in the second variation of the Lagrangian.

Gauss and Hertz

Other extremal principles of classical mechanics have been formulated, such as Gauss' principle of least constraint and its corollary, Hertz's principle of least curvature.

Apparent Teleology

The mathematical equivalence of the differential equations of motion and their integral counterpart has important philosophical implications. The differential equations are statements about quantities localised to a single point in space or single moment of time. For example, Newton's second law F = ma states that the instantaneous force F applied to a mass m produces an acceleration a at the same instant. By contrast, the action principle is not localised to a point; rather, it involves integrals over an interval of time and (for fields) an extended region of space. Moreover, in the usual formulation of classical action principles, the initial and final states of the system are fixed, e.g.,

> *Given that the particle begins at position x_1 at time t_1 and ends at position x_2 at time t_2, the physical trajectory that connects these two endpoints is an extremum of the action integral.*

In particular, the fixing of the *final* state appears to give the action principle a teleological character which has been controversial historically. However, some critics maintain this apparent teleology occurs because of the way in which the question was asked. By specifying some but not all aspects of both the initial and final conditions (the positions but not the velocities) we are making some inferences about the initial conditions from the final conditions, and it is this "backward" inference that can be seen as a teleological causal influence.

The speculative fiction writer, Ted Chiang, has a story, *Story of Your Life*, that contains visual depictions of Fermat's Principle along with a discussion of its teleological dimension. Keith Devlin's *The Math Instinct* contains a chapter, "Elvis the Welsh Corgi Who Can Do Calculus" that discusses the calculus "embedded" in some animals as they solve the "least time" problem in actual situations.

Maupertuis' Principle

In classical mechanics, Maupertuis' principle (named after Pierre Louis Maupertuis) is an integral equation that determines the *path* followed by a physical system without specifying the time parameterization of that path. It is a special case of the more generally stated principle of least action. More precisely, it is a formulation of the equations of motion for a physical system not as differential equations, but as an *integral equation*, using the calculus of variations.

Jacobi's Formulation

For many systems, the kinetic energy T is quadratic in the generalised velocities $\dot{\mathbf{q}}$

$$T = \frac{1}{2}\frac{d\mathbf{q}}{dt}\cdot\mathbf{M}\cdot\frac{d\mathbf{q}}{dt}$$

although the mass tensor $\mathbf{M}$ may be a complicated function of the generalised coordinates $\mathbf{q}$. For such systems, a simple relation relates the kinetic energy, the generalised momenta and the generalised velocities

$$2T = \mathbf{p}\cdot\dot{\mathbf{q}}$$

provided that the potential energy $V(\mathbf{q})$ does not involve the generalised velocities. By defining a normalised distance or metric ds in the space of generalised coordinates

$$ds^2 = d\mathbf{q}\cdot\mathbf{M}\cdot d\mathbf{q}$$

one may immediately recognise the mass tensor as a metric tensor. The kinetic energy may be written in a massless form

$$T = \frac{1}{2}\left(\frac{ds}{dt}\right)^2$$

or, equivalently,

$$2Tdt = \mathbf{p}\cdot d\mathbf{q} = \sqrt{2T}\ ds$$

Hence, the abbreviated action can be written

$$\mathcal{S}_0 \stackrel{\text{def}}{=} \int \mathbf{p} \cdot d\mathbf{q} = \int ds \sqrt{2} \sqrt{E_{tot} - V(\mathbf{q})}$$

since the kinetic energy $T = E_{tot} - V(\mathbf{q})$ equals the (constant) total energy E_{tot} minus the potential energy $V(\mathbf{q})$. In particular, if the potential energy is a constant, then Jacobi's principle reduces to minimising the path length $s = \int ds$ in the space of the generalised coordinates, which is equivalent to Hertz's principle of least curvature.

Kirchhoff's Diffraction Formula

Kirchhoff's diffraction formula can be used to model the propagation of light in a wide range of configurations, either analytically or using numerical modelling. It gives an expression for the wave disturbance when a monochromatic spherical wave passes through an opening in an opaque screen. The equation is derived by making several approximations to the Kirchhoff integral theorem which uses Green's theorem to derive the solution to the homogeneous wave equation.

Derivation of Kirchhoff's Diffraction Formula

Kirchhoff's integral theorem, sometimes referred to as the Fresnel-Kirchhoff integral theorem, usesGreen's theorem to derive the solution to the homogeneous wave equation at an arbitrary point P in terms of the values of the solution of the wave equation and its first order derivative at all points on an arbitrary surface which encloses P. The solution provided by the integral theorem for a monochromatic source is:

$$U(P) = \frac{1}{4\pi} \int_S \left[U \frac{\partial}{\partial n} \left(\frac{e^{iks}}{s} \right) - \frac{e^{iks}}{s} \frac{\partial U}{\partial n} \right] dS$$

where U is the complex amplitude of the disturbance at the surface, k is the wavenumber and s is the distance from P to the surface.

Point Source

Consider a monochromatic point source at P_0 which illuminates an aperture in a screen. The energy of the wave emitted by a point source falls off as the inverse square of the distance travelled, so the amplitude falls off as the inverse of the distance. The complex amplitude of the disturbance at a distance r is given by

$$U(r) = \frac{ae^{ikr}}{r}$$

where a represents the magnitude of the disturbance at the point source.

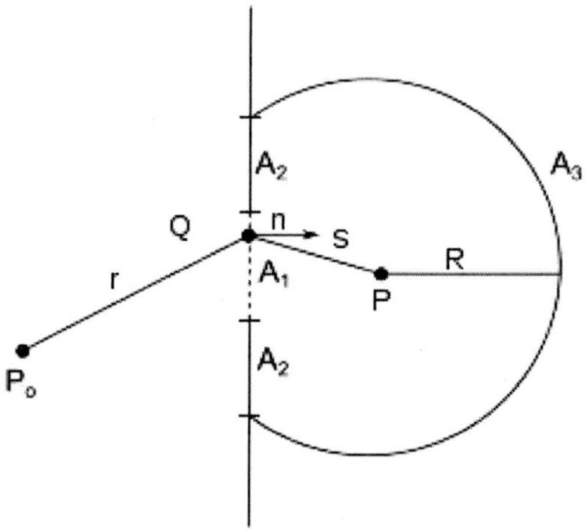

Figure: Geometrical arrangement used in deriving Kirchhhoff's diffraction formula

The disturbance at a point P can be found by applying the integral theorem to the closed surface which is formed by the intersection of a sphere of radius R with the screen. The integration is performed over the areas A_1, A_2 and A_3 giving

$$U(P)=\frac{1}{4\pi}\left[\int_{A_1}+\int_{A_2}+\int_{A_3}\left[U\frac{\partial}{\partial n}\left(\frac{e^{iks}}{s}\right)-\frac{e^{iks}}{s}\frac{\partial U}{\partial n}\right]\right]dS$$

To solve the equation, it is assumed that the values of U and $\partial U/\partial n$ in the area A_1 are the same as when the screen is not present, giving:

$$U_{A_1}=\frac{ae^{ikr}}{r}$$

$$\frac{\partial U_{A_1}}{\partial n}=\frac{ae^{ikr}}{r}\left[ik-\frac{1}{r}\right]\cos(n,r)$$

at P_0, r is the length P_0Q, and (n,r) is the angle between P_0Q and the normal to the aperture.

Kirchoff assumes that the values of U and $\partial U/\partial n$ in A_2 are zero. This implies that U and $\partial U/\partial n$ are discontinuous at the edge of the aperture. This is not the case, and this is one of the approximation used in deriving the equation. These assumptions are sometimes referred to as Kirchhoff's boundary conditions. The contribution from A_3 to the integral is also assumed to be zero. This can be justified by making the assumption that the source starts to radiate at a particular time, and then by making R large enough, so that when the disturbance at P is being considered, no contributions from A_3 will have arrived there. Such a wave is no longer monochromatic, since a monochromatic wave must exist at all times, but that assumption is not necessary, and a more formal argument avoiding its use has been derived.

We have

$$\frac{\partial}{\partial n}\left(\frac{e^{iks}}{s}\right)=\frac{e^{iks}}{s}\left[ik-\frac{1}{s}\right]\cos(n,s)$$

where (n,s) is the angle between the normal to the aperture and QP.

Finally, the terms $1/r$ and $1/s$ are assumed to be negligible compared with k, since r and s are generally much greater than $2\pi/k$ which is equal to the wavelength. Thus, the integral above, which represents the complex amplitude at P, becomes:

$$U(P)=-\frac{ia}{2\lambda}\int_S\frac{e^{ik(r+s)}}{rs}[\cos(n,r)-\cos(n,s)]dS$$

This is the Kirchhoff or Fresnel–Kirchhoff diffraction formula.

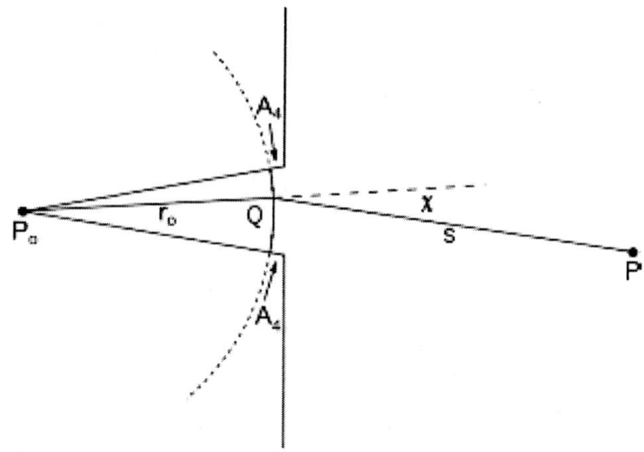

Figure : Geometric arrangement used to express Kirchhoff's formula in a form similar to Huygens–Fresnel

Equivalence to Huygens–Fresnel Equation

The Huygens–Fresnel principle can be derived by integrating over a different closed surface. The area A_1 above is replaced by a wavefront from P_0 which almost fills the aperture, and a portion of a cone with a vertex at P_0 which is labelled A_4 in the diagram. If the radius of curvature of the wave is large enough, the contribution from

A_4 can be neglected. We also have $\chi = \pi - (r_0,s)$ whereχ is as defined in Huygens–Fresnel principlea nd $\cos(n,r) = 1$. The complex amplitude of the wavefront at r_0 is given by:

$$U(r_0) = \frac{ae^{ikr_0}}{r_0}$$

The diffraction formula becomes;

$$U(P) = -\frac{i}{2\lambda}\frac{ae^{ikr_0}}{r_0}\int_S \frac{e^{iks}}{s}(1+\cos\chi)dS$$

This is the Kirchhoff's diffraction formula which contains parameters which had to be arbitrarily assigned in the derivation of the Huygens–Fresnel equation.

Extended Source

Assume that the aperture is illuminated by an extended source wave. The complex amplitude at the aperture is given by $U_0(r)$

It is assumed, as before, that the values of U and $\partial U/\partial n$ in the area A_1 are the same as when the screen is not present, that the values of U and $\partial U/\partial n$ in A_2 are zero (Kirchhoff's boundary conditions) and that the contribution from A_3 to the integral are also zero. It is also assumed that $1/s$ is negligible compared with k. We then have

$$U(P) = -\frac{1}{4\pi}\int_{A_1}\frac{e^{iks}}{s}\left[ikU_0(r)\cos(n,s) + \frac{\partial U_0(r)}{\partial n}\right]dS$$

This is the most general form of the Kirchhoff diffraction formula. To solve this equation for an extended source, an additional integration would be required to sum the contributions made by the individual points in the source. If, however, we assume that the light from the source at each point in the aperture has a well-defined direction, which is the case of the distance between the source and the aperture is significantly greater than the wavelength, then we can write

$$U_0(r) \approx a(r)e^{-ikr}$$

where $a(r)$ is the magnitude of the disturbance at the point r in the aperture. We then have:

$$\frac{\partial U_0(r)}{\partial n} = -ika(r)\cos(n,r)$$

We then have:

$$U(P) = -\frac{i}{2\lambda}\int_S a(r)\frac{e^{iks}}{s}[\cos(n,r) - \cos(n,s)]dS\ .$$

Fraunhofer and Fresnel Diffraction Equations

In spite of the various approximations which were made in arriving at the formula, it is adequate to describe the majority of problems in instrumental optics. This is mainly because the wavelength of light is much smaller than the dimensions of any obstacles encountered. Analytical solutions are not possible for most configurations but the Fresnel diffraction equation and Fraunhofer diffractionequation, which are approximations of Kirchhoff's formula for the near field and far field, can be applied to a very wide range of optical systems.

One of the important assumptions made in arriving at the Kirchhoff diffraction formula is that r,s are significantly greater than λ. A further approximation can be made which significantly simplifies the equation further: this is that the distances P_0Q and QP are much greater than the dimensions of the aperture. This allows one to make two further approximations:

cos(n,r)-cos(n,s) is replaced with 2cos β where β is the angle between P_0P and the normal to the aperture. The factor 1/rs is replaced with 1/r's' where r' , s' are the distances from P_0 and P to the origin, which is located in the aperture. The complex amplitude then becomes:

$$U(P) = -\frac{ia\cos\beta}{\lambda r's'}\int_S e^{ik(r+s)}ds$$

We assume that the aperture lies in the *yz* plane, and the co-ordinates of P_0, P and Q (a general point in the aperture) are (x_0, y_0, x_0), (x,y,z) and $(0,y',z')$ respectively. We then have:

$$r^2 = (x_0 - x')^2 + (y_0 - y')^2 + z_0^2$$

$$s^2 = (x - x')^2 + (y - y')^2 + z^2$$

$$r'^2 = x_0^2 + y_0^2 + z_0^2$$

$$s'^2 = x^2 + y^2 + z^2$$

We can express *r* and *s* as follows:

$$r = r'\left[1 - \frac{2(x_0x' + y_0y')}{r'} + \frac{x'^2 + y'^2}{r'}\right]^{1/2}$$

$$s = s'\left[1 - \frac{2(xx' + yy')}{s'} + \frac{x'^2 + y'^2}{s'}\right]^{1/2}$$

These can be expanded as power series:

$$r = r'\left[\begin{array}{l} 1 - \frac{1}{2r'}[2(x_0x' + y_0y') + (x'^2 + y'^2)] \\ + \frac{1}{2r'}[2(x_0x' + y_0y') + (x'^2 + y'^2)]^2 + \dots\dots \end{array}\right]$$

$$s = s'\left[\begin{array}{l} 1 - \frac{1}{2s'}[2(xx' + yy') + (x'^2 + y'^2)] \\ + \frac{1}{2s'}[2(xx' + yy') + (x'^2 + y'^2)]^2 + \dots\dots \end{array}\right]$$

The complex amplitude at P can now be expressed as:

$$U(P) = -\frac{i\cos\beta}{\lambda}\frac{ae^{ik(r'+s')}}{r's'}\int_S e^{ikf(x',y')}dx'dy'$$

where $f(x',y')$ includes all the terms in the expressions above for s and r apart from the first term in each expression and can be written in the form:

$$f(x'y') = c_1x' + c_2y' + c_3x'^2 + c_4y'^2 + c_5x'y' \dots\dots$$

where the c are constants.

Fraunhofer Diffraction

If all the terms in $f(x',y')$ can be neglected except for the terms in x' and y', we have the Fraunhofer diffraction equation. If the direction cosines of P_0Q and PQ are;

$$\begin{aligned} l_0 &= -x_0 / r' \\ m_0 &= -y_0 / r' \\ l &= x / s' \\ m &= y / s' \end{aligned}$$

the Fraunhofer diffraction equation is then

$$U(P) = C\int_S e^{ik[(l_0 - l)x' + (m - m_0)y']}dx'dy'$$

where C is a constant. This can also be written in the form:

$$U(P) = C\int_S e^{i(\mathbf{k_0} - \mathbf{k})\cdot\mathbf{r}'}dr'$$

where k_0 and k are the wave vectors of the waves travelling from P_0 to the aperture and from the aperture to P respectively, and r' is a point in the aperture. If the point source is replaced by an extended source whose complex amplitude at the aperture is given by U_0(r') then the Fraunhofer diffraction equation is:

$$U(P) \propto \int_S a_0(\mathbf{r}')e^{i(\mathbf{k}_0-\mathbf{k})\cdot\mathbf{r}'}dr'$$

where a_0(r') is, as before, the magnitude of the disturbance at the aperture.

Fresnel Diffraction

When the quadratic terms cannot be neglected but all higher order terms can, the equation becomes the Fresnel diffraction equation.

Assumptions Made

The diffraction equation is arrived at by applying Green's theorem to the solution of the wave equation. The approximations made in deriving the Kirchhoff, Fraunhofer and Fresnel equations are summarised in the following sections.

Kirchhoff Diffraction Equation

- U and $\partial U/\partial n$ are discontinuous at the boundaries of the aperture
- r and s are much greater than λ.

Fraunhofer Diffraction Equations

In addition to the approximations made in deriving the Kirchhoff equation, it is assumed that

- *r,s* are significantly greater than the size of the aperture
- Second and higher order terms in the expression *f(x',y')* can be neglected.

Fresnel Diffraction Equation

In addition to the approximations used in deriving the Kirchhoff equation, it is assumed that

- *r,s* are significantly greater than the size of the aperture
- Third and higher order terms in the expression *f(x',y')* can be neglected.

Green's Function

In mathematics, a Green's function is a type of function used to solve inhomogeneous differential equations subject to specific initial conditions or boundary conditions. Under many-body theory, the term is also used in physics, specifically in quantum field theory, electrodynamics and statistical field theory, to refer to various types of correlation functions, even those that do not fit the mathematical definition.

Green's functions are named after the British mathematician George Green, who first developed the concept in the 1830s. In the modern study of linear partial differential equations, Green's functions are studied largely from the point of view of fundamental solutions instead.

Definition and Uses

A Green's function, G(x, s), of a linear differential operator L = L(x) acting on distributions over a subset of the Euclidean space R^n, at a point s, is any solution of

$$LG(x,s) = \delta(x - s) \qquad (1)$$

where δ is the Dirac delta function. This property of a Green's function can be exploited to solve differential equations of the form

$$Lu(x) = f(x) \qquad (2)$$

If the kernel of *L* is non-trivial, then the Green's function is not unique. However, in practice, some combination of symmetry, boundary conditions and/or other externally imposed criteria will give a unique Green's function.

Also, Green's functions in general are distributions, not necessarily proper functions.

Green's functions are also a useful tool in solving wave equations, diffusion equations, and in quantum mechanics, where the Green's function of the Hamiltonian is a key concept, with important links to the concept of density of states. As a side note, the Green's function as used in physics is usually defined with the opposite sign; that is,

$LG(x,s) = - \delta(x$ ” $s)$.

This definition does not significantly change any of the properties of the Green's function.

If the operator is translation invariant, that is when *L* has constant coefficients with respect to *x*, then the Green's function can be taken to be a convolution operator, that is,

$G(x,s) = G(x - s)$.

In this case, the Green's function is the same as the impulse response of linear time-invariant system theory.

Motivation

Loosely speaking, if such a function G can be found for the operator L, then if we multiply the equation (1) for the Green's function by f(s), and then perform an integration in the s variable, we obtain;

$$\int LG(x,s)f(s)ds = \int \delta(x-s)f(s)ds = f(x)$$

The right hand side is now given by the equation (2) to be equal to $L\ u(x)$, thus:

$$Lu(x) = \int LG(x,s)f(s)ds$$

Because the operator $L = L(x)$ is linear and acts on the variable x alone (not on the variable of integration s), we can take the operator L outside of the integration on the right hand side, obtaining;

$$Lu(x) = L\left(\int G(x,s)f(s)ds\right)$$

And this suggests;

$$u(x) = \int G(x,s)f(s)ds \qquad (3)$$

Thus, we can obtain the function $u(x)$ through knowledge of the Green's function in equation (1), and the source term on the right hand side in equation (2). This process relies upon the linearity of the operator L.

In other words, the solution of equation (2), $u(x)$, can be determined by the integration given in equation (3). Although $f(x)$ is known, this integration cannot be performed unless G is also known. The problem now lies in finding the Green's function G that satisfies equation (1). For this reason, the Green's function is also sometimes called the fundamental solution associated to the operator L.

Not every operator L admits a Green's function. A Green's function can also be thought of as a right inverse of L. Aside from the difficulties of finding a Green's function for a particular operator, the integral in equation (3), may be quite difficult to evaluate.

However the method gives a theoretically exact result.

This can be thought of as an expansion of f according to a Dirac delta function basis (projecting f over ä$(x - s)$) and a superposition of the solution on each projection. Such an integral equation is known as a Fredholm integral equation, the study of which constitutes Fredholm theory.

Green's Functions for Solving Inhomogeneous Boundary Value Problems

The primary use of Green's functions in mathematics is to solve non-homogeneous boundary value problems. In modern theoretical physics, Green's functions are also usually used as propagators in Feynman diagrams (and the phrase Green's function is often used for any correlation function).

Framework

Let L be the Sturm–Liouville operator, a linear differential operator of the form

$$L = \frac{d}{dx}\left[p(x)\frac{d}{dx}\right] + q(x)$$

and let D be the boundary conditions operator •

$$Du = \begin{cases} \alpha_1 u'(0) + \beta_1 u(0) \\ \alpha_2 u'(l) + \beta_2 u(l) \end{cases}$$

Let $f(x)$ be a continuous function in [0,l]. We shall also suppose that the problem

$$\begin{aligned} Lu &= f \\ Du &= 0 \end{aligned}$$

is regular (i.e., only the trivial solution exists for the homogeneous problem).

Theorem

There is one and only one solution $u(x)$ which satisfies

$$\begin{aligned} Lu &= f \\ Du &= 0 \end{aligned}$$

and it is given by

$$u(x) = \int_0^{\ell} f(s)G(x,s)ds$$

where $G(x,s)$ is a Green's function satisfying the following conditions:

1. $G(x,s)$ is continuous in x and s
2. For $x \neq s$, $LG(x,s) = 0$
3. For $s \neq 0$, $DG(x,s) = 0$
4. Derivative "jump": $G'(s_{+0},s) - G'(s_{-0},s) = 1/p(s)$
5. Symmetry: $G(x, s) = G(s, x)$

Finding Green's Functions

Eigenvalue Expansions

If a differential operator L admits a set of eigenvectors $\psi_n(x)$ (i.e., a set of functions ψ_n and scalars λν συχη that $L\psi_n = \lambda_n\psi_n$)) that is complete, then it is possible to construct a Green's function from these eigenvectors and eigenvalues. Complete means that the set of functions $\{\Psi_n\}$ satisfies the following completeness relation:

$$\delta(x - x') = \sum_{n=0}^{\infty} \Psi_n^{\dagger}(x)\Psi_n(x')$$

Then the following holds:

$$G(x,x') = \sum_{n=0}^{\infty} \frac{\Psi_n^{\dagger}(x)\Psi_n(x')}{\lambda_n}$$

where $\dagger$ represents complex conjugation.

Applying the operator L to each side of this equation results in the completeness relation, which was assumed true.

The general study of the Green's function written in the above form, and its relationship to the function spaces formed by the eigenvectors, is known as Fredholm theory.

Green's Functions for the Laplacian

Green's functions for linear differential operators involving the Laplacian may be readily put to use using the second of Green's identities.

To derive Green's theorem, begin with the divergence theorem (otherwise known as Gauss's theorem):

$$\int_V \nabla\cdot\vec{A}\, dV = \int_S \vec{A}\cdot d\hat{\sigma}$$

Let $\vec{A} = \phi\nabla\psi - \psi\nabla\phi$ and substitute into Gauss' law. Compute $\nabla\cdot\vec{A}$ and apply the chain rule for the ∇ operator:

$$\begin{aligned}\nabla\cdot\vec{A} &= \nabla\cdot(\phi\nabla\psi - \psi\nabla\phi)\\ &= (\nabla\phi)\cdot(\nabla\psi) + \phi\nabla^2\psi - (\nabla\phi)\cdot(\nabla\psi) - \psi\nabla^2\phi\\ &= \phi\nabla^2\psi - \psi\nabla^2\phi\end{aligned}$$

Plugging this into the divergence theorem produces Green's theorem:

$$\int_V (\phi\nabla^2\psi - \psi\nabla^2\phi)dV = \int_S (\phi\nabla\psi - \psi\nabla\phi)\cdot d\hat{\sigma}$$

Suppose that the linear differential operator L is the Laplacian, ∇^2, and that there is a Green's function G for the Laplacian. The defining property of the Green's function still holds:

$$LG(x,x') = \nabla^2 G(x,x') = \delta(x - x')$$

Let $\psi = G$ in Green's theorem. Then:

$$\int_V \left[\phi(x')\delta(x-x') - G(x,x')\nabla^2\phi(x')\right] d^3x'$$

$$[6pt] \quad = \int_S \left[\phi(x')\nabla' G(x,x') - G(x,x')\nabla'\phi(x')\right]\cdot d\hat{\sigma}'$$

Using this expression, it is possible to solve Laplace's equation $\nabla^2\phi(x) = 0$ or Poisson's equation $\nabla^2\phi(x) = -\rho(x)$, subject to either Neumann or Dirichlet boundary conditions. In other words, we can solve for $\varphi(x)$ everywhere inside a volume where either (1) the value of $\varphi(x)$ is specified on the bounding surface of the volume (Dirichlet boundary conditions), or (2) the normal derivative of $\varphi(x)$ is specified on the bounding surface (Neumann boundary conditions).

Suppose the problem is to solve for $\varphi(x)$ inside the region. Then the integral

$$\varphi\varphi\int_V \phi(x')\delta(x-x')\, d^3x'$$

reduces to simply $\varphi(x)$ due to the defining property of the Dirac delta function and we have:

$$\phi(x) = \int_V G(x,x')\rho(x')\, d^3x' + \int_S \left[\phi(x')\nabla' G(x,x') - G(x,x')\nabla'\phi(x')\right]\cdot d\hat{\sigma}'$$

This form expresses the well-known property of harmonic functions that if the value or normal derivative is known on a bounding surface, then the value of the function inside the volume is known everywhere.

In electrostatics, $\varphi(x)$ is interpreted as the electric potential, $\rho(x)$ as electric charge density, and the normal derivative $\nabla\phi(x')\cdot d\hat{\sigma}'$ as the normal component of the electric field.

If the problem is to solve a Dirichlet boundary value problem, the Green's function should be chosen such that $G(x,x')$ vanishes when either x or x' is on the bounding surface. Thus only one of the two terms in the surface integral remains. If the problem is to solve a Neumann boundary value problem, the Green's function is chosen such that its normal derivative vanishes on the bounding surface, as it would seems to be the most logical choice. However, application of Gauss's theorem to the differential equation defining the Green's function yields

$$\int_S \nabla' G(x,x')\cdot d\hat{\sigma}' = \int_V \nabla'^2 G(x,x')d^3x' = \int_V \delta(x-x')d^3x' = 1$$

meaning the normal derivative of $G(x,x')$ cannot vanish on the surface, because it must integrate to 1 on the surface. The simplest form the normal derivative can take is that of a constant, namely 1/ S, where S is the surface area of the surface. The surface term in the solution becomes

$$\int_S \phi(x')\nabla' G(x,x')\cdot d\hat{\sigma}' = \langle\phi\rangle_S$$

where $\langle\phi\rangle_S$ is the average value of the potential on the surface. This number is not known in general, but is often unimportant, as the goal is often to obtain the electric field given by the gradient of the potential, rather than the potential itself.

With no boundary conditions, the Green's function for the Laplacian (Green's function for the three-variable Laplace equation) is:

$$G(x,x') = \frac{1}{|x - x'|}$$

Supposing that the bounding surface goes out to infinity, and plugging in this expression for the Green's function, this gives the familiar expression for electric potential in terms of electric charge density (in the CGS unit system) as

$$\phi(x) = \int_V \frac{\rho(x')}{|x - x'|} d^3x'$$

Example

Given the problem

$$Lu = u'' + u = f(x)$$

$$u(0) = 0, \quad u\left(\frac{\pi}{2}\right) = 0$$

Find the Green's function.

First step: The Green's function for the linear operator at hand is defined as the solution to

$$g''(x,s) + g(x,s) = \delta(x - s).$$

If $x \neq s$, then the delta function gives zero, and the general solution is

$$g(x,s) = c_1\cos x + c_2\sin x.$$

For $x < s$, the boundary condition at $x = 0$ implies

$$g(0,s) = c_1\cdot 1 + c_2\cdot 0 = 0, \quad c_1 = 0$$

The equation of $g\left(\frac{\pi}{2}, s\right) = 0$ is skipped because $x \neq \frac{\pi}{2}$ if $x < s$

and $s \neq \frac{\pi}{2}$

For $x > s$, the boundary condition at $x = \pi/2$ implies

$$g\left(\frac{\pi}{2}, s\right) = c_3 \cdot 0 + c_4 \cdot 1 = 0, \quad c_4 = 0$$

The equation of $g(0,s) = 0$ is skipped for similar reasons.

To summarize the results thus far:

$$g(x,s) = \begin{cases} c_2 \sin x, & \text{for } x < s \\ c_3 \cos x, & \text{for } s < x \end{cases}$$

Second step: The next task is to determine c_2 and c_3.

Ensuring continuity in the Green's function at $x = s$ implies

$$c_2 \sin s = c_3 \cos s$$

One can also ensure proper discontinuity in the first derivative by integrating the defining differential equation from $x = s - \epsilon$ to $x = s + \epsilon$ and taking the limit as ϵ goes to zero:

$$c_3 \cdot [-\sin s] - c_2 \cos s = 1$$

The two (dis)continuity equations can be solved for c_2 and c_3 to obtain

$$c_2 = -\cos s \quad ; \quad c_3 = -\sin s$$

So the Green's function for this problem is:

$$g(x,s) = \begin{cases} -\cos s \sin x, & x < s \\ -\sin s \cos x, & s < x \end{cases}$$

Further Examples

- Let $n = 1$ and let the subset be all of R. Let L be *d*/*dx*. Then, the Heaviyide step function $H(x - x_0)$ is a Green's function of L at x_0.
- Let $n = 2$ and let the subset be the quarter-plane { (*x*, *y*) : *x*, *y* e" 0 } and L be the Laplacian. Also, assume a Dirichlet boundary condition is imposed at $x = 0$ and a Neumann boundary condition is imposed at $y = 0$. Then the Green's function is

$$G(x, y, x_0, y_0)$$

$$[6pt] \quad = \frac{1}{2\pi}\left[\ln\sqrt{(x-x_0)^2 + (y-y_0)^2} - \ln\sqrt{(x+x_0)^2 + (y-y_0)^2}\right]$$

$$[6pt] \quad + \frac{1}{2\pi}\left[\ln\sqrt{(x-x_0)^2 + (y+y_0)^2} - \ln\sqrt{(x+x_0)^2 + (y+y_0)^2}\right]$$

Green's Theorem

In mathematics, Green's theorem gives the relationship between aline integral around a simple closed curve C and a double integral over the plane region D bounded by C. It is the two-dimensional special case of the more general Stokes' theorem, and is named after British mathematician George Green.

Let C be a positively oriented, piecewise smooth, simple closed curve in the plane $\mathbb{R}^2$, and let D be the region bounded by C. If L and$M$ ae functions of (x, y) defined on an open region containing D and have continuous partial derivatives there, then

$$\oint_C (L\mathrm{d}x + M\mathrm{d}y) = \iint_D \left(\frac{\partial M}{\partial x} - \frac{\partial L}{\partial y}\right)\mathrm{d}x\,\mathrm{d}y.$$

For positive orientation, an arrow pointing in the counterclockwise direction may be drawn in the small circle in the integral symbol.

In physics, Green's theorem is mostly used to solve two-dimensional flow integrals, stating that the sum of fluid outflows at any point inside a volume is equal to the total outflow summed about an enclosing area.

In plane geometry, and in particular, area surveying, Green's theorem can be used to determine the area and centroid of plane figures solely by integrating over the perimeter.

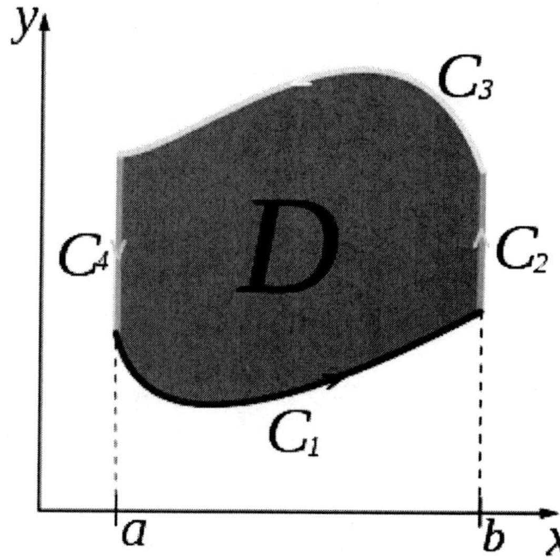

***Figure :** If D is a simple region with its boundary consisting of the curves C_1, C_2, C_3, C_4, Green's theorem can be demonstrated.*

Proof when D is a Simple Region

The following is a proof of the theorem for the simplified area D, a type I region where C_2 and C_4 are vertical lines. A similar proof exists for when D is a type II region where *C1 and* C_3 are straight lines. The general case can be deduced from this special case by approximating the domain D by a union of simple domains.

If it can be shown that

$$\int_C L dx = \iint_D \left(-\frac{\partial L}{\partial y}\right) dA \qquad (1)$$

and

$$\int_C M dy = \iint_D \left(\frac{\partial M}{\partial x}\right) dA \qquad (2)$$

are true, then Green's theorem is proven in the first case.

Define the type I region D as pictured on the right by:

$$D = \{(x, y) \mid a \le x \le b, g_1(x) \le y \le g_2(x)\}$$

where g_1 and g_2 are continuous functions on $[a, b]$. Compute the double integral in (1):

$$\begin{aligned}\iint_D \frac{\partial L}{\partial y} dA &= \int_a^b \int_{g_1(x)}^{g_2(x)} \frac{\partial L}{\partial y}(x, y) dy dx \\ &= \int_a^b \left\{L(x, g_2(x)) - L(x, g_1(x))\right\} dx \qquad (3)\end{aligned}$$

Now compute the line integral in (1). C can be rewritten as the union of four curves: C_1, C_2, C_3, C_4.

With C_1, use the parametric equations: $x = x$, $y = g_1(x)$, a d" x d" b. Then

$$\int_{C_1} L(x, y) dx = \int_a^b L(x, g_1(x)) dx$$

With C_3, use the parametric equations: $x = x$, $y = g_2(x)$, a d" x d" b. Then

$$\int_{C_3} L(x, y) dx = -\int_{-C_3} L(x, y) dx = -\int_a^b L(x, g_2(x)) dx$$

The integral over C_3 is negated because it goes in the negative direction from b to a, as C is oriented positively (counterclockwise). On C_2 and C_4, x remains constant, meaning

$$\int_{C_4} L(x, y) dx = \int_{C_2} L(x, y) dx = 0$$

Therefore,

$$\int_C L dx = \int_{C_1} L(x,y)dx + \int_{C_2} L(x,y)dx + \int_{C_3} L(x,y)dx + \int_{C_4} L(x,y)dx$$
$$= -\int_a^b L(x,g_2(x))dx + \int_a^b L(x,g_1(x))dx \qquad (4)$$

Combining (3) with (4), we get (1). Similar computations give (2).

Relationship to the Stokes Theorem

Green's theorem is a special case of Stokes' theorem, when applied to a region in the xy-plane: We can augment the two-dimensional field into a three-dimensional field with a z component that is always 0. Write F for the vector-valued function $\mathbf{F} = (L, M, 0)$. Start with the left side of Green's theorem:

$$\oint_C (Ldx + Mdy) = \oint_C (L, M, 0)\cdot(dx, dy, dz) = \oint_C \mathbf{F}\cdot d\mathbf{r}$$

Then by Stokes' Theorem:

$$\oint_C \mathbf{F}\cdot d\mathbf{r} = \iint_S \nabla\times\mathbf{F}\cdot\hat{\mathbf{n}}dS$$

The surface S is just the region in the plane D, with the unit normals $\hat{\mathbf{n}}$ pointing up (in the positive z direction) to match the "positive orientation" definitions for both theorems.

The expression inside the integral becomes

$$\nabla\times\mathbf{F}\cdot\hat{\mathbf{n}} = \left[\left(\frac{\partial 0}{\partial y} - \frac{\partial M}{\partial z}\right)\mathbf{i} + \left(\frac{\partial L}{\partial z} - \frac{\partial 0}{\partial x}\right)\mathbf{j} + \left(\frac{\partial M}{\partial x} - \frac{\partial L}{\partial y}\right)\mathbf{k}\right]\cdot\mathbf{k} = \left(\frac{\partial M}{\partial x} - \frac{\partial L}{\partial y}\right)$$

Thus we get the right side of Green's theorem

$$\iint_S \nabla\times\mathbf{F}\cdot\hat{\mathbf{n}}dS = \iint_D \left(\frac{\partial M}{\partial x} - \frac{\partial L}{\partial y}\right)dA$$

Relationship to the Divergence Theorem

Considering only two-dimensional vector fields, Green's theorem is equivalent to the following two-dimensional version of the divergence theorem:

$$\iint_D (\nabla\cdot\mathbf{F})dA = \oint_C \mathbf{F}\cdot\hat{\mathbf{n}}ds,$$

where $\hat{\mathbf{n}}$ is the outward-pointing unit normal vector on the boundary.

To see this, consider the unit normal in the right side of the equation.

Since $d\mathbf{r} = (dx, dy)$ is a vector pointing tangential along a curve, and the curve C is the positively-oriented (i.e. counterclockwise) curve along the boundary, an outward normal would be a vector which points 90° to the right, which would be $(dy, -dx)$. The length of this vector is $\sqrt{dx^2 + dy^2} = ds$. So $\hat{\mathbf{n}}ds = (dy, -dx)$.

Now let the components of $\mathbf{F} = (P, Q)$. Then the right hand side becomes

$$\int_C \mathbf{F}\cdot\hat{\mathbf{n}}ds = \int_C Pdy - Qdx$$

which by Green's theorem becomes

$$\int_C -Qdx + Pdy = \iint_D \left(\frac{\partial P}{\partial x} + \frac{\partial Q}{\partial y}\right)dA = \iint_D (\nabla\cdot\mathbf{F})dA.$$

Area Calculation

Green's theorem can be used to compute area by line integral. The area of D is given by:

$$A = \iint_D \mathrm{d}A.$$

Provided we choose L and M such that:

$$\frac{\partial M}{\partial x} - \frac{\partial L}{\partial y} = 1$$

Then the area is given by:

$$A = \oint_C (L\mathrm{d}x + M\mathrm{d}y)$$

Possible formulas for the area of D include:

$$A = \oint_C x\mathrm{d}y = -\oint_C y\mathrm{d}x = \tfrac{1}{2}\oint_C (-y\mathrm{d}x + x\mathrm{d}y)$$

Double-slit Experiment

The double-slit experiment, sometimes called Young's experiment, is a demonstration that matter and energy can display characteristics of both waves and particles. In the basic version of the experiment, a coherent light source such as a laser beam illuminates a thin plate pierced by two parallel slits, and the light passing through the slits is observed on a screen behind the plate. The wave nature of light causes the light waves passing through the two slits to interfere, producing bright and dark bands on the screen — a result that would not be expected if light consisted strictly of particles. However, at the screen, the light is always found to be absorbed as though it were composed of discrete particles or photons. This establishes the principle known as wave–particle duality.

Overview

If light consisted strictly of ordinary or classical particles, and these particles were fired in a straight line through a slit and allowed to strike a screen on the other side, we would expect to see a pattern corresponding to the size and shape of the slit. However, when this "single-slit experiment" is actually performed, the pattern on the screen is a diffraction pattern, a fairly narrow central band with dimmer bands parallel to it on each side.

Similarly, if light consisted strictly of classical particles and we illuminated two parallel slits, the expected pattern on the screen would simply be the sum of the two single-slit patterns. In actuality, however, the pattern becomes wider and much more detailed, including a series of light and dark bands.

When Thomas Young first demonstrated this phenomenon, it indicated that light consists of waves, as the distribution of brightness can be explained by the alternately additive and subtractive interference of wavefronts.

Young's experiment played a vital part in the acceptance of the wave theory of light in the early 1800s, vanquishing the corpuscular theory of light proposed by Isaac Newton, which had been the accepted model of light propagation in the 17th and 18th centuries. However, the later discovery of the photoelectric effect demonstrated that under different circumstances, light can behave as if it is composed of discrete particles. These seemingly contradictory discoveries made it necessary to go beyond classical physics and take the quantumnature of light into account.

A double-slit experiment was not performed with anything other than light until 1961, when Clauss Jφnsson of the University of Tübingen performed it with electrons. In 2002, Jφnsson's double-slit experiment was voted "the most beautiful experiment" by readers of Physics World.

In 1999, objects large enough to be seen under a microscope — buckyball molecules (diametre about 0.7 nm, nearly half a million times larger than a proton) — were found to exhibit wave-like interference. The double-slit experiment (and its variations) has become a classic thought experiment for its clarity in expressing the central puzzles of quantum mechanics. Richard Feynman was fond of saying that all of quantum mechanics can be gleaned from carefully thinking through the implications of this single experiment.

Variations of the Experiment

Interference of Individual Particles

An important version of this experiment involves single particles (or waves — for consistency, they are called particles here). Sending particles through a double-slit apparatus one at a time results in single particles appearing on the screen, as expected. Remarkably, however, an interference pattern emerges when these particles are allowed to build up one by one. For example, when a laboratory apparatus was developed that could reliably fire one electron at a time through the double slit, the emergence of an interference pattern suggested that *each electron was interfering with itself*, and therefore in some sense the electron *had to be going through both slits at once* — an idea that contradicts our everyday experience of discrete objects. This phenomenon has also been shown to occur with atoms and even some molecules, including buckyballs. So experiments with electrons add confirmatory evidence to the view of Dirac that electrons, protons, neutrons, and even larger entities that are ordinarily called particles nevertheless have their own wave nature and even their own specific frequencies.

This experimental fact is highly reproducible, and the mathematics of quantum mechanics allows us to predict the exact probability of an electron striking the screen at any particular point. However, the electrons do not arrive at the screen in any predictable order. In other words, knowing where all the previous electrons appeared on the screen and in what order tells us nothing about where any future electron will hit, even though the probabilities at specific points can be calculated. Thus, we have the appearance of a seemingly causeless selection event in a highly orderly and predictable formulation of the interference pattern. This has prompted some theorists to look for additional determinants or "hidden variables" in the system that, were they to become known, would account for the location of each individual impact with the target.

A low-intensity double-slit experiment was first performed by Taylor in 1909, by reducing the level of incident light until photon emission/absorption events were mostly nonoverlapping. The appearance of interference built up from individual photons could be explained by understanding that a single photon has its own wavefront that passes through both slits, and that the single photon will show up on the detector screen according to the net probability values resulting from the co-incidence of the two probability waves coming by way of the two slits. Note that it is not the *probabilities* of photons appearing at various points along the detection screen that add or cancel, but the *amplitudes*. If there is a cancellation of waves at some point, that does not mean that a photon disappears; it only means that the probability of a photon's appearing at that point will decrease, and the probability that it will appear somewhere else increases.

A detailed treatment of the mathematics of double-slit interference in the context of quantum mechanics is given in the article on Englert–Greenberger duality.

With Particle Detectors at the Slits

The double-slit apparatus can be modified by adding particle detectors positioned at the slits. This enables the experimenter to find the position of a particle not when it impacts the screen, but rather, when it passes through the double-slit — did it go through only one of the slits, as a particle would be expected to do, or through both, as a wave would be expected to do? Numerous experiments have shown, however, that any modification of the apparatus that can determine which slit a particle passes through reduces the visibility of interference at the screen, thereby illustrating the complementarity principle: that light (and electrons, etc.) can behave as either particles or waves, but not both at the same time. An experiment performed in 1987 produced results that demonstrated that information could be obtained regarding which path a particle had taken, without destroying the interference altogether. This showed the effect of measurements that disturbed the particles in transit to a lesser degree and thereby influenced the interference pattern only to a comparable extent.

There are many methods to determine whether a photon passed through a slit, for instance by placing an atom at the position of each slit. Interesting experiments of this latter kind have been performed with photons and with neutrons.

Delayed Choice and Quantum Eraser Variations

The delayed-choice experiment and the quantum eraser are sophisticated variations of the double-slit with particle detectors placed not at the slits but elsewhere in the apparatus. The first demonstrates that extracting "which path" information *after* a particle passes through the slits can seem to retroactively alter its previous behaviour at the slits. The second demonstrates that wave behaviour can be restored by erasing or otherwise making permanently unavailable the "which path" information.

Other Variations

In 1967 Pfleegor and Mandel demonstrated two-source interference using two separate lasers as light sources.

It was shown experimentally in 1972 that in a double-slit system where only one slit was open at any time, interference was nonetheless observed provided the path difference was such that the detected photon could have come from either slit. The experimental conditions were such that the photon density in the system was much less than unity.

The experiment has been performed with particles as large as C60(Buckminsterfullerene).

Classical Wave-optics Formulation

Much of the behaviour of light can be modelled using classical wave theory. The Huygens–Fresnel principle is one such model; it states that each point on a wavefront generates a secondary spherical wavelet, and that the disturbance at any subsequent point can be found by summing the contributions of the individual wavelets at that point. This summation needs to take into account the phase as well as the amplitude of the individual wavelets. It should be noted that only the intensity of a light field can be measured – this is proportional to the square of the amplitude.

In the double-slit experiment, the two slits are illuminated by a single laser beam. If the width of the slits is small enough (less than the wavelength of the laser light), the slits diffract the light into cylindrical waves. These two cylindrical wavefronts are superimposed, and the amplitude, and therefore the intensity, at any point in the combined wavefronts depends on both the magnitude and the phase of the two wavefronts. The difference in phase between the two waves is determined by the difference in the distance travelled by the two waves.

If the viewing distance is large compared with the separation of the slits (the far field), the phase difference can be found using the geometry shown in the figure below right. The path difference between two waves travelling at an angle θ is given by: $d\sin\theta \approx d\theta$

When the two waves are in phase, i.e. the path difference is equal to an integral number of wavelengths, the summed amplitude, and therefore the summed intensity is maximum, and when they are in anti-phase, i.e. the path difference is equal to half a wavelength, one and a half wavelengths, etc., then the two waves cancel and the summed intensity is zero. This effect is known as interference. The interference fringe maxima occur at angles

$$d\theta_n = n\lambda, n = 0,1,2,\ldots$$

where λ is the wavelength of the light. The angular spacing of the fringes is θ_f is given by

$$\theta_f \approx \lambda / d$$

The spacing of the fringes at a distance *z* from the slits is given by

$$w = z\theta_f = z\lambda / d$$

For example, if two slits are separated by 0.5mm, and are illuminated with a 0.6μm wavelength laser, the spacing of the fringes viewed at a distance of 1m will be 1.2mm.

If the width of the slits *b* is greater than the wavelength, the Fraunhofer diffraction equation gives the intensity of the diffracted light as:

$$I(\theta) \quad \propto \cos^2\left[\frac{\pi d \sin\theta}{\lambda}\right]\mathrm{sinc}^2\left[\frac{\pi b sin\theta}{\lambda}\right]$$

This is illustrated in the figure above, where the first pattern is the diffraction pattern of a single slit, given by the sinc function in this equation, and the second figure shows the combined intensity of the light diffracted from the two slits, where the cos function represent the fine structure, and the coarser structure represents diffraction by the individual slits as described by the sinc function. Similar calculations for the near field can be done using the Fresnel diffraction equation. As the plane of observation gets closer to the plane in which the slits are located, the diffraction patterns associated with each slit decrease in size, so that the area in which interference occurs is reduced, and may vanish altogether when there is no overlap in the two diffracted patterns.

Interpretations of the Experiment

Like the Schrφdinger's cat thought experiment, the double-slit experiment is often used to highlight the differences and similarities between the various interpretations of quantum mechanics.

Copenhagen Interpretation

The Copenhagen interpretation is a consensus among some of the pioneers in the field of quantum mechanics that it is undesirable to posit anything that goes beyond the mathematical formulae and the kinds of physical apparatus and reactions that enable us to gain some knowledge of what goes on at the atomic scale.

One of the mathematical constructs that enables experimenters to predict very accurately certain experimental results is sometimes called a probability wave. In its mathematical form it is analogous to the description of a physical wave, but its "crests" and "troughs" indicate levels of probability for the occurrence of certain phenomena (e.g., a spark of light at a certain point on a detector screen) that can be observed in the macro world of ordinary human experience.

The probability "wave" can be said to "pass through space" because the probability values that one can compute from its mathematical representation are dependent on time.

One cannot speak of the location of any particle such as a photon between the time it is emitted and the time it is detected simply because in order to say that something is located somewhere at a certain time one has to detect it. The requirement for the eventual appearance of an interference pattern is that particles be emitted, and that there be a screen with at least two distinct paths for the particle to take from the emitter to the detection screen. Experiments observe nothing whatsoever between the time of emission of the particle and its arrival at the detection screen. If a ray tracing is then made as if a light wave (as understood in classical physics) is wide enough to take both paths, then that ray tracing will accurately predict the appearance of maxima and minima on the detector screen when many particles pass through the apparatus and gradually "paint" the expected interference pattern.

Path-integral Formulation

The Copenhagen interpretation is similar to the path integral formulation of quantum mechanics provided by Feynman. The path integral formulation replaces the classical notion of a single, unique trajectory for a system, with a sum over all possible trajectories. The trajectories are added together by using functional integration.

Each path is considered equally likely, and thus contributes the same amount. However, the phase of this contribution at any given point along the path is determined by the action along the path:

$$A_{path}(x,t) = e^{iS(x,t)}$$

All these contributions are then added together, and the magnitude of the final result is squared, to get the probability distribution for the position of a particle:

$$p(x,t) \propto \left| \int_{allpaths(x,t)} e^{iS(x,t)} \right|^2$$

As is always the case when calculating probability, the results must then be normalised:

$$\int_{x \in X} p(x,t)dx = 1$$

To summarize, the probability distribution of the outcome is the normalised square of the norm of the superposition, over all paths from the point of origin to the final point, of waves propagating proportionally to the action along each path.

The differences in the cumulative action along the different paths (and thus the relative phases of the contributions) produces the interference pattern observed by the double-slit experiment. Feynman stressed that his formulation is merely a mathematical description, not an attempt to describe a real process that we cannot measure.

Relational Interpretation

According to the relational interpretation of quantum mechanics, first proposed by Carlo Rovelli, observations such as those in the double-slit experiment result specifically from the interaction between the observer (measuring device) and the object being observed (physically interacted with), not any absolute property possessed by the object. In the case of an electron, if it is initially "observed" at a particular slit, then the observer–particle (photon–electron) interaction includes information about the electron's position.

This partially constrains the particle's eventual location at the screen. If it is "observed" (measured with a photon) not at a particular slit but rather at the screen, then there is no "which path" information as part of the interaction, so the electron's "observed" position on the screen is determined strictly by its probability function.

This makes the resulting pattern on the screen the same as if each individual electron had passed through both slits. It has also been suggested that space and distance themselves are relational, and that an electron can appear to be in "two places at once" – for example, at both slits – because its spatial relations to particular points on the screen remain identical from both slit locations.

Optical sine Theorem

In optics, the optical sine theorem states that the products of the index, height, and sine of the slope angle of a ray in object space and its corresponding ray in image space are equal. That is:

$$n_o y_o \alpha_o = n_i y_i \alpha_i$$

Optical Theorem

In physics, the optical theorem is a general law of wave scattering theory, which relates the forward scattering amplitude to the total cross section of the scatterer. It is usually written in the form

$$\sigma_{\text{tot}} = \frac{4\pi}{k}\text{Im} f(0),$$

where f(0) is the scattering amplitude with an angle of zero, that is, the amplitude of the wave scattered to the centre of a distant screen. Because the optical theorem is derived using only conservation of energy, or in quantum mechanics from conservation of probability, the optical theorem is widely applicable and, in quantum mechanics, σ_{tot} includes both elastic and inelastic scattering. Note that the above form is for an incident plane wave; a more general form discovered by Werner Heisenberg can be written

$$\text{Im} f(\hat{\mathbf{k}}', \hat{\mathbf{k}}) = \frac{k}{4\pi}\int f(\hat{\mathbf{k}}', \hat{\mathbf{k}}'') f(\hat{\mathbf{k}}'', \hat{\mathbf{k}}) d\hat{\mathbf{k}}''.$$

Notice that as a natural consequence of the optical theorem, an object that scatters any light at all ought to have a nonzero forward scattering amplitude, and so a bright central spot.

History

The optical theorem was originally discovered independently by Sellmeier and Lord Rayleigh in 1871. Lord Rayleigh recognised the forward scattering amplitude in terms of the index of refraction as

$$n = 1 + 2\pi \frac{N f(0)}{k^2},$$

which he used in a study of the colour and polarisation of the sky. The equation was later extended to quantum scattering theory by several individuals, and came to be known as the Bohr–Peierls–Placzek relation after a 1939 publication. It was first referred to as the Optical Theorem in print in 1955 by Hans Bethe and Frederic de Hoffmann, after it had been known as a “well known theorem of optics” for some time.

Derivation

The theorem can be derived rather directly from a treatment of a scalar wave. If a plane wave is incident on an object, then the wave amplitude a great distance away from the scatterer is given approximately by

$$\psi(\mathbf{r}) \approx e^{ikz} + f(\theta)\frac{e^{ikr}}{r}.$$

All higher terms, when squared, vanish more quickly than $1/r^2$, and so are negligible a great distance away. Notice that for large values of z and small angles the binomial theorem gives us

$$r = \sqrt{x^2 + y^2 + z^2} \approx z + \frac{x^2 + y^2}{2z}.$$

We would now like to use the fact that the intensity is proportional to the square of the amplitude ø. Approximating the r in the denominator as z, we have

$$|\psi|^2 = |e^{ikz} + \frac{f(\theta)}{z}e^{ikz}e^{ik(x^2+y^2)/2z}|^2$$

$$= 1 + \frac{f(\theta)}{z}e^{ik(x^2+y^2)/2z} + \frac{f^*(\theta)}{z}e^{-ik(x^2+y^2)/2z} + \frac{|f(\theta)|^2}{z^2}.$$

If we drop the $1/z^2$ term and use the fact that $A + A^* = 2\mathrm{Re}\,A$ we have

$$|\psi|^2 \approx 1 + 2\mathrm{Re}\left(\frac{f(\theta)}{z}e^{ik(x^2+y^2)/2z}\right).$$

Now suppose we integrate over a screen in the *x-y* plane, at a distance which is small enough for the small angle approximations to be appropriate, but large enough that we can integrate the intensity from $-\infty$ to ∞ with negligible error. In optics, this is equivalent to including many fringes of the diffraction pattern. To further simplify matters, let's approximate $f(\theta) = f(0)$. We quickly obtain

$$\int |\psi|^2\, da \approx A + 2\mathrm{Re}\left(\frac{f(0)}{z}\int_{-\infty}^{\infty} e^{ikx^2/2z}dx \int_{-\infty}^{\infty} e^{iky^2/2z}dy\right)$$

where A is the area of the surface integrated over. The exponentials can be treated as Gaussians and so

$$\int |\psi|^2\, da = A + 2\mathrm{Re}\left(\frac{f(0)}{z}\frac{i2z\pi}{k}\right)$$

$$= A - \frac{4\pi}{k} \operatorname{Im} f(0),$$

which is just the probability of reaching the screen if none were scattered, lessened by an amount $(4\pi / k)\operatorname{Im} f(0)$, which is therefore the effective scattering cross section of the scatterer.

Reflection Coefficient

he reflection coefficient is used in physics and electrical engineering when wave propagation in a medium containing discontinuities is considered. A reflection coefficient describes either the amplitudeor the intensity of a reflected wave relative to an incident wave. The reflection coefficient is closely related to the *transmission coefficient.*

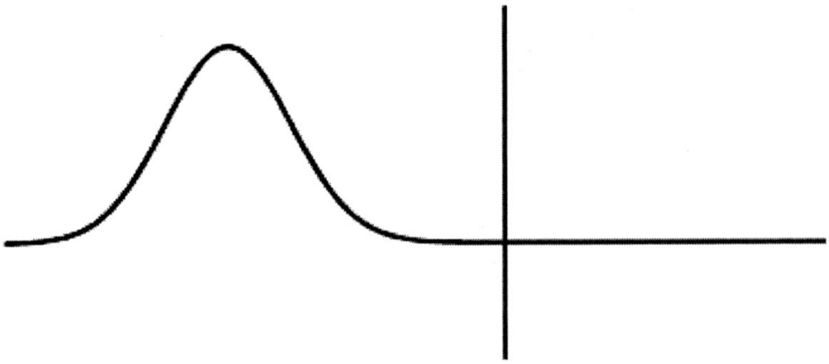

Figure: A wave experiences partial transmittance and partial reflectance when the medium through which it travels suddenly changes.

Different specialities have different applications for the term.

Telecommunications

In telecommunications, the reflection coefficient is the ratio of the amplitude of the reflected wave to the amplitude of the incident wave. In particular, at a discontinuity in a transmission line, it is the complex ratio of the electric field strength of the reflected wave (E^-) to that of the incident wave (E^+). This is typically represented with a Γ (capital gamma) and can be written as:

$$\Gamma = \frac{E^-}{E^+}$$

The reflection coefficient may also be established using other field or circuit quantities.

The reflection coefficient can be given by the equations below, where Z_S is the impedance toward the source, Z_L is the impedance toward the load:

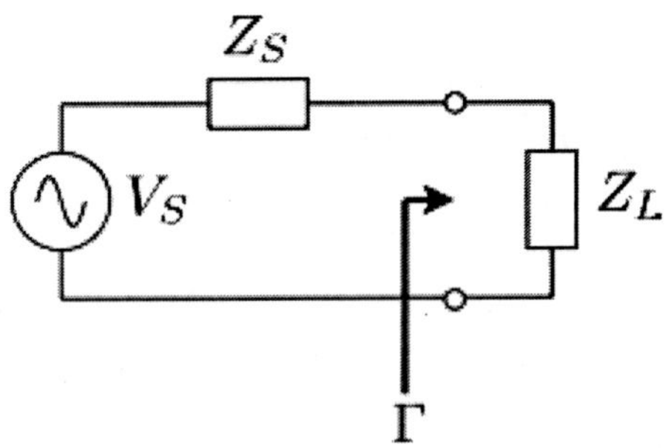

Simple circuit configuration showing measurement location of reflection coefficient.

$$\Gamma = \frac{Z_L - Z_S}{Z_L + Z_S}$$

Notice that a negative reflection coefficient means that the reflected wave receives a 180°, or π, phase shift.

The absolute magnitude (designated by vertical bars) of the reflection coefficient can be calculated from the standing wave ratio, *SWR*:

$$|\Gamma| = \frac{SWR - 1}{SWR + 1}$$

The reflection coefficient is displayed graphically using aSmith chart.

Seismology

Reflection coefficient is used in feeder testing for reliability of medium.

Optics and Microwaves

In optics, intensity and amplitude reflection coefficients are used. Typically, the former are represented by a capital R, while the latter are represented by a lower-case r.

Semipermeable Membranes

The reflection coefficient in semipermeable membranes relates to how such a membrane can reflect solute particles from passing through. A value of zero results in all particles passing through. A value of one is such that no particle can pass. It is used in the Starling equation.

Power Cables in high-power Electrical Drives

The reflection coefficient characterises an amplitude of voltage reflections between motor and inverter

Refractive Index

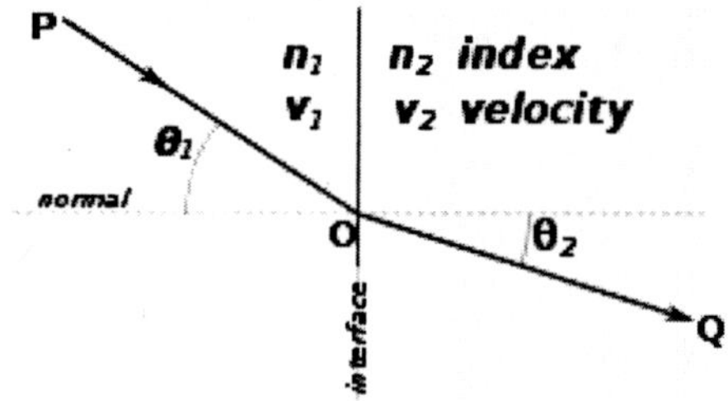

Figure : In optics the refractive index or index of refraction of a substance or medium is a measure of the speed of light in that medium. It is expressed as a ratio of the speed of light in vacuum relative to that in the considered medium. This can be written mathematically as:

n = speed of light in a vacuum/speed of light in medium.

For example, the refractive index of water is 1.33, meaning that light travels 1.33 times as fast in vacuum as it does in water.

As light moves from a medium, such as air, water, or glass, into another it may change its propagation direction in proportion to the change in refractive index. This refraction is governed by Snell's law, and is illustrated in the figure to the right. Refractive index of materials varies with the wavelength of light. This is called dispersion and results in a slightly different refractive index for each colour.

The wavelength λ of light in a material is determined by the refractive index according to $\lambda = \lambda_0 / n$, where λ_0 is the wavelength of the light in vacuum. Brewster's angle, the critical angle for total internal reflection, and the reflectivity of a surface is also affected by the refractive index. These material parameters can be calculated using the Fresnel equations.

The concept of refractive index is widely used within the full electromagnetic spectrum, from x-rays to radio waves. It can also be used with wave phenomena other than light, e.g. sound. In this case the speed of sound is used instead of that of light and a reference medium other than vacuum must be chosen.

Definitions

The refractive index, n, of a medium is defined as the ratio of the speed, c, of a wave phenomenon such as light or sound in a reference medium to the phase speed, v_p, of the wave in the medium in question:

$$n = \frac{c}{v_p}.$$

It is most commonly used in the context of light with vacuum as a reference medium, although historically other reference media (e.g. air at a standardized pressure and temperature) have been common. It is usually given the symbol *n*. In the case of light, it equals

$$n = \sqrt{\epsilon_r \mu_r},$$

where ϵ_r is the material's relative permittivity, and μ_r is its relative permeability. For most naturally occurring materials, μ_r is very close to 1 at optical frequencies, therefore *n* is approximately $\sqrt{\epsilon_r}$. Contrary to a widespread misconception, the real part of a complex *n* may be less than one, depending upon the material and wavelength.This has practical technical applications, such as effective mirrors for X-rays based on total external reflection.

The phase speed is defined as the rate at which the crests of the waveform propagate; that is, the rate at which the phase of the waveform is moving.

The *group speed* is the rate at which the *envelope* of the waveform is propagating; that is, the rate of variation of the amplitude of the waveform.

Provided the waveform is not distorted significantly during propagation, it is the group speed that represents the rate at which information (and energy) may be transmitted by the wave (for example, the speed at which a pulse of light travels down an optical fibre). For the analytic properties constraining the unequal phase and group speeds in dispersive media, refer to dispersion (optics). Another common definition of the refractive index comes from the refraction of a light ray entering a medium. The refractive index is the ratio of the sines of the angles of incidence θ_1 and refraction θ_2 as light passes into the medium or mathematically

$$n = \frac{\sin\theta_1}{\sin\theta_2}.$$

The angles are measured to the normal of the surface. This definition is based on Snell's law and is equivalent to the definition above if the light enters from the reference medium (normally vacuum).

A closely related quantity is *refractivity*, which in atmospheric applications is denoted *N* and defined as $N = 10^6(n - 1)$. The 10^6 factor is used because for air, *n* deviates from unity at most a few parts per thousand.

Typical Values

For visible light most transparent media have refractive indices between 1 and 2. A few examples are given in the table to the right. These values are measured at the yellow doublet sodium D-line, with a wavelength of 589 nanometres, as is conventionally done. Gases at atmospheric pressure have refractive indices close to 1 because of their low density. Most plastics have refractive indices in the range from 1.3 to 1.7, but some high-refractive-index polymers can have a value as high as 1.76. For infrared light refractive indices can be considerably higher. Germanium is transparent in this region and has a refractive index of about 4, making it an important material for infrared optics.

Refractive Index Below 1

A widespread misconception is that since, according to the theory of relativity, nothing can travel faster than the speed of light in vacuum, the refractive index cannot be lower than 1. This is erroneous since the refractive index measures the phase velocity of light, which does not carry energy or information, the two things limited in propagation speed. The phase velocity is the speed at which the crests of the wave move and can be faster than the speed of light in vacuum, and thereby give a refractive index below 1. This can occur close to resonance frequencies, inplasmas, and for x-rays. In the x-ray regime the refractive indices are lower than but very close to 1 (exceptions close to some resonance frequencies). As an example, water has a refractive index of 1"2.6×10$^{"7}$ at a photon energy of 30 keV (0.04 nm wavelength).

Negative Refractive Index

Recent research has also demonstrated the existence of the negative refractive index, which can occur if permittivity and permeability have simultaneous negative values. This can be achieved with periodically constructed metamaterials. The resulting negative refraction (i.e., a reversal of Snell's law) offers the possibility of the superlens and other exotic phenomena.

Microscopic Explanation

At the microscale, an electromagnetic wave's phase speed is slowed in a material because the electric field creates a disturbance in the charges of each atom (primarily the electrons) proportional to the electric susceptibility of the medium. (Similarly, the magnetic field creates a disturbance proportional to the magnetic susceptibility.) As the electromagnetic fields oscillate in the wave, the charges in the material will be "shaken" back and forth at the same frequency. The charges thus radiate their own electromagnetic wave that is at the same frequency, but usually with a phase delay, as the charges may move out of phase with the force driving them. The light wave travelling in the medium is the macroscopic superposition (sum) of all such contributions in the material: The original wave plus the waves radiated by all the moving charges. This wave is typically a wave with the same frequency but shorter wavelength than the original, leading to a slowing of the wave's phase speed. Most of the radiation from oscillating material charges will modify the incoming wave, changing its velocity. However, some net energy will be radiated in other directions or even at other frequencies. Depending on the relative phase of the original driving wave and the waves radiated by the charge motion, there are several possibilities:

- If the electrons emit a light wave which is 90° out of phase with the light wave shaking them, it will cause the total light wave to travel more slowly. This is the normal refraction of transparent materials like glass or water, and corresponds to a refractive index which is real and greater than 1.
- If the electrons emit a light wave which is 270° out of phase with the light wave shaking them, it will cause the total light wave to travel more quickly. This is called "anomalous refraction", and is observed close to absorption lines, with X-rays, and in some microwave systems. It corresponds to a refractive index less than 1. (Even though the phase velocity of light is greater than the speed of light in vacuum *c*, the signal velocity is not, as discussed above). If the response is sufficiently strong and out-of-phase, the result is negative refractive index discussed below.
- If the electrons emit a light wave which is 180° out of phase with the light wave shaking them, it will destructively interfere with the original light to reduce the total light intensity. This is light absorption in opaque materials and corresponds to an imaginary refractive index.
- If the electrons emit a light wave which is in phase with the light wave shaking them, it will amplify the light wave. This is rare, but occurs in lasers due to stimulated emission. It corresponds to an imaginary index of refraction, with the opposite sign as absorption.

For most materials at visible-light frequencies, the phase is somewhere between 90° and 180°, corresponding to a combination of both refraction and absorption.

Dispersion

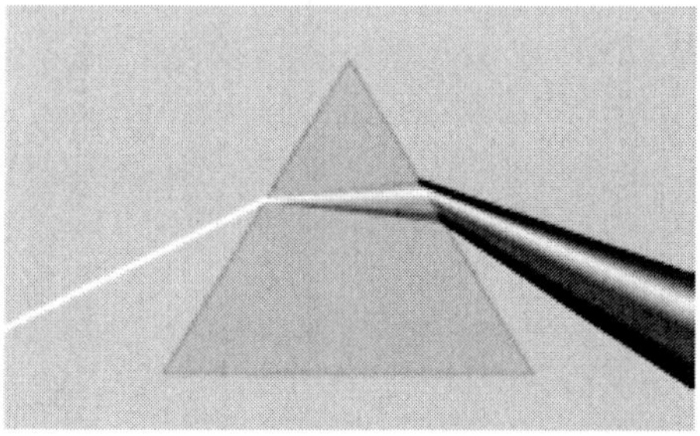

Figure: In a prism dispersion causes different colours to refract at different angles, splitting white light into a rainbow of colours.

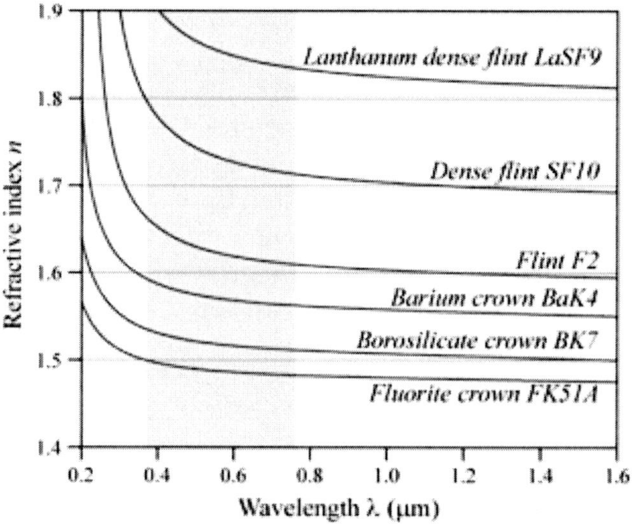

Figure : The variation of refractive index with wavelength for various glasses.

The refractive index of materials varies with the wavelength (and frequency) of light. This is called dispersion and causes prisms to divide white light into its constituent spectral colours, and explains how rainbows are formed. As the refractive index varies with wavelength, according to Snell's law, so will the refraction angle as light goes from one material to another. This makes different colours go in different directions. Dispersion also causes the focal length of lenses to be wavelength dependent. This is a type of chromatic aberration, which often needs to be corrected for in imaging systems. In regions of the spectrum where the material does not absorb, the refractive index tends to decrease with increasing wavelength, and thus increase with frequency. This is called normal dispersion, in contrast to anomalous dispersion, where the refractive index increases with wavelength. For visible light normal dispersion means that the refractive index is higher for blue light than for red.

For optics in the visual range the amount of dispersion of a lens material is often quantified by the Abbe number $V = \frac{n_{\text{yellow}} - 1}{n_{\text{blue}} - n_{\text{red}}}$. For a more accurate description of the wavelength dependence of the refractive index the Sellmeier equation can be used. It is an empirical formula that works well in describing dispersion. *Sellmeier coefficients* are often quoted instead of the refractive index in tables.

Because of dispersion, it is usually important to specify the vacuum wavelength at which a refractive index is measured. Typically, this is done at various well-defined spectral emission lines; for example, nD is the refractive index at the Fraunhofer "D" line, the centre of the yellow sodium double emission at 589.29 nm wavelength.

Complex Index of Refraction and Absorption

When light passes through a medium, some part of it will always be absorbed. This can be conveniently taken into account by defining a complex index of refraction,

$$\tilde{n} = n + i\kappa.$$

Here, the real part of the refractive index n indicates the phase speed, while the imaginary part κindicates the amount of absorption loss when the electromagnetic wave propagates through the material.

That κ corresponds to absorption can be seen by inserting this refractive index into the expression for electric field of a plane electromagnetic wave travelling in the z-direction. We can do this by relating the wave number to the refractive index through $k = \frac{2\pi n}{\lambda_0}$, with λ_0 being the vacuum wavelength. With complex wave number $\tilde{k}$ and refractive index $n + i\kappa$ this can be inserted into the plane wave expression as

$$\mathbf{E}(z,t) = \text{Re}(\mathbf{E}_0 e^{i(\tilde{k}z - \omega t)}) = \text{Re}(\mathbf{E}_0 e^{i(2\pi(n+i\kappa)z/\lambda_0 - \omega t)}) = e^{-2\pi\kappa z/\lambda_0}\text{Re}(\mathbf{E}_0 e^{i(kz-\omega t)}).$$

Here we see that κ gives an exponential decay, as expected from the Beer–Lambert law.

κ is often called the extinction coefficient in physics although this has a different definition within chemistry. Both n and κ are dependent on the frequency. In most circumstances $\kappa > 0$ (light is absorbed) or $\kappa = 0$ (light travels forever without loss). In special situations, especially in the gain medium of lasers, it is also possible that $\kappa < 0$, corresponding to an amplification of the light.

An alternative convention uses $\tilde{n} = n - i\kappa$ instead of $\tilde{n} = n + i\kappa$, but where $\kappa > 0$ still corresponds to loss. Therefore these two conventions are inconsistent and should not be confused. The difference is related to defining sinusoidal time dependence as Re($e^{-i\grave{u}t}$) versus Re($e^{+i\grave{u}t}$).

Dielectric loss and non-zero DC conductivity in materials cause absorption. Good dielectric materials such as glass have extremely low DC conductivity, and at low frequencies the dielectric loss is also negligible, resulting in almost no absorption (κ H" 0). However, at higher frequencies (such as visible light), dielectric loss may increase absorption significantly, reducing the material's transparency to these frequencies.

The real and imaginary parts of the complex refractive index are related through the Kramers–Kronig relations. For example, one can determine a material's full complex refractive index as a function of wavelength from an absorption spectrum of the material. For X-ray and extreme ultraviolet radiation the complex refractive index deviates only slightly from unity and usually has a real part smaller than 1. It is therefore normally written as $\tilde{n} = 1 - \delta + i\beta$ (or $\tilde{n} = 1 - \delta - i\beta$).

Relations to Other Quantities

Refraction

When light moves from one medium to another as in the figure to the right, it changes direction, i.e. it is refracted. If it goes from a medium with refractive index n_1 to one with refractive index n_2, with an incidence angle to the surface normal of θ_1, the transmission angle θ_2 can be calculated from Snell's law:

$$n_1 \sin\theta_1 = n_2 \sin\theta_2.$$

If there is no angle θ_2 fulfilling Snell's law, i.e.

$$\frac{n_1}{n_2}\sin\theta_1 > 1,$$

the light cannot be transmitted and will instead undergototal internal reflection.

Reflectivity

Apart from the transmitted light there is also a reflected part. The reflection angle is equal to the incidence angle, and the amount of light that is reflected is determined by the reflectivity of the surface. The reflectivity can be calculated from the refractive index and the incidence angle with the Fresnel equations, which for normal incidence reduces to

$$R = \left(\frac{n_1 - n_2}{n_1 + n_2} \right)^2 .$$

For common glass in air, $n_1 = 1$ and $n_2 = 1.5$, and thus about 4% of the incident power is reflected. At other incidence angles the reflectivity will also depend on the polarisation of the incoming light. At a certain angle called Brewster's angle, p-polarised light (light with the electric field in the plane of incidence) will be totally transmitted. Brewster's angle can be calculated from the two refractive indices of the interface as

$$\theta_B = \arctan\left(\frac{n_2}{n_1} \right).$$

Lenses

The focal length of a lens is determined by its refractive index n and the radii of curvature R_1 and R_2 of its surfaces. The power of a thin lens in air is given by the Lensmaker's formula:

$$\frac{1}{f} = (n-1)\left(\frac{1}{R_1} - \frac{1}{R_2} \right).$$

Dielectric Constant

The frequency dependent dielectric constant is simply the square of the (complex) refractive index in a non-magnetic medium (one with a relative permeability of unity). The refractive index is used for optics in Fresnel equations and Snell's law; while the dielectric constant is used in Maxwell's equations and electronics. Where $\tilde{\epsilon}$ is the complex dielectric constant with real and imaginary parts ϵ_1 ϵ_2, and n and κ are the real and imaginary parts of the refractive index, all functions of frequency:

$$\tilde{\epsilon} = \epsilon_1 + i\epsilon_2 = (n + i\kappa)^2 .$$

Conversion between refractive index and dielectric constant is done by:

$$\epsilon_1 = n^2 - \kappa^2$$

$$\epsilon_2 = 2n\kappa$$

$$n = \sqrt{\frac{\sqrt{\epsilon_1^2 + \epsilon_2^2} + \epsilon_1}{2}}$$

$$\kappa = \sqrt{\frac{\sqrt{\epsilon_1^2 + \epsilon_2^2} - \epsilon_1}{2}}.$$

Density

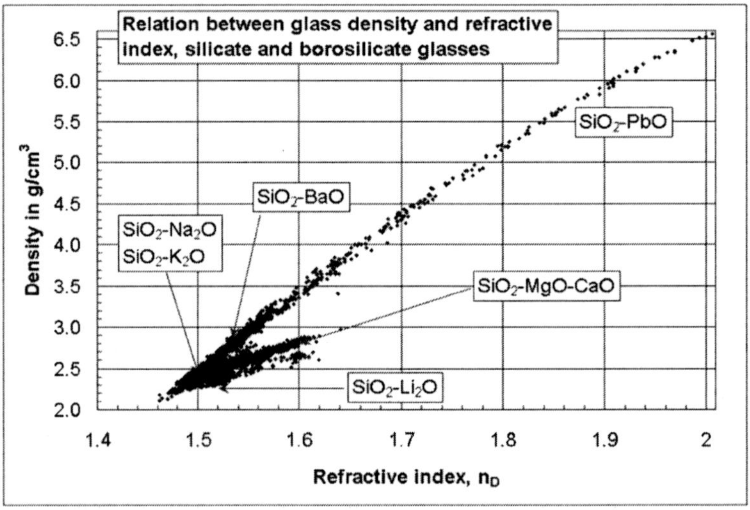

Figure : Relation between the refractive index and the density of silicate and borosilicate glasses.

In general, the refractive index of a glass increases with its density. However, there does not exist an overall linear relation between the refractive index and the density for all silicate and borosilicate glasses. A relatively high refractive index and low density can be obtained with glasses containing light metal oxides such as Li_2O and MgO, while the opposite trend is observed with glasses containing PbO and BaO as seen in the diagram at the right.

Group Index

Sometimes, a "group speed refractive index", usually called the *group index* is defined:

$$n_g = \frac{c}{v_g}$$

where v_g is the group velocity. This value should not be confused with n, which is always defined with respect to the phase velocity. When the dispersion is small, the group velocity can be linked to the phase velocity by the relation

$$v_g = v - \lambda \frac{dv}{d\lambda}.$$

In this case the group index can thus be written in terms of the wavelength dependence of the refractive index as

$$n_g = \frac{n}{1 + \frac{\lambda}{n}\frac{dn}{d\lambda}},$$

where λ is the wavelength in the medium.

When the refractive index of a medium is known as a function of the vacuum wavelength (instead of the wavelength in the medium), the corresponding expressions for the group velocity and index are (for all values of dispersion)

$$v_g = c\left(n - \lambda_0 \frac{dn}{d\lambda_0}\right)^{-1}.$$

$$n_g = n - \lambda_0 \frac{dn}{d\lambda_0},$$

where λ_0 is the wavelength in vacuum.

Other Relations

As shown in the Fizeau experiment, when light is transmitted through a moving medium, its speed relative to a stationary observer is:

$$V = \frac{c}{n} + v\left(1 - \frac{1}{n^2}\right)$$

The refractive index of a substance can be related to its polarisability with the Lorentz–Lorenz equation or to the molar refractivities of its constituents by the Gladstone–Dale relation.

Birefringence

In some materials the refractive index depends on the polarisation and propagation direction of the light. This is called birefringence or optical anisotropy. In the simplest form, uniaxial birefringence, there is only one special direction in the material. This axis is known as the optical axis of the material. Light with linear polarisation perpendicular to this axis will experience an *ordinary refractive* index n_o while light polarised in parallel will experience an *extraordinary* refractive index n_e. The birefringence of the material is the difference between these indices of refraction, $\Delta n = n_e - n_o$. Light propagating in the direction of the optical axis will not be affected by the birefringence since the refractive index will be *no independent* of polarisation. For other propagation directions the light will split into two linearly polarised beams. For light travelling perpendicularly to the optical axis the beams will have the same direction. This can be used to change the polarisation direction of linearly polarised light or to convert between linear, circular and elliptical polarisations with waveplates.

Many crystals are naturally birefringent, but isotropic materials such as plastics and glass can also often be made birefringent by introducing a prefered direction through e.g. an external force or electric field. This can be utilised in the determination of stresses in structures using photoelasticity. The birefringent material is then placed between crossed polarisers. A change in birefringence will alter the polarisation and thereby the fraction of light that is transmitted through the second polariser.

In the more general case of trirefringent materials described by the field of crystal optics, the *dielectric constant* is a rank-2 tensor (a 3 by 3 matrix). In this case the propagation of light cannot simply be described by refractive indices except for polarisations along principal axes.

Nonlinearity

The strong electric field of high intensity light (such as output of a laser) may cause a medium's refractive index to vary as the light passes through it, giving rise to nonlinear optics. If the index varies quadratically with the field (linearly with the intensity), it is called the optical Kerr effect and causes phenomena such as self-focusing and self-phase modulation. If the index varies linearly with the field (which is only possible in materials that do not possess inversion symmetry), it is known as the Pockels effect.

Inhomogeneity

If the refractive index of a medium is not constant, but varies gradually with position, the material is known as a gradient-index medium and is described by gradient index optics. Light travelling through such a medium can be bent or focused, and this effect can be exploited to produce lenses, some optical fibres and other devices. Some common mirages are caused by a spatially-varying refractive index of air.

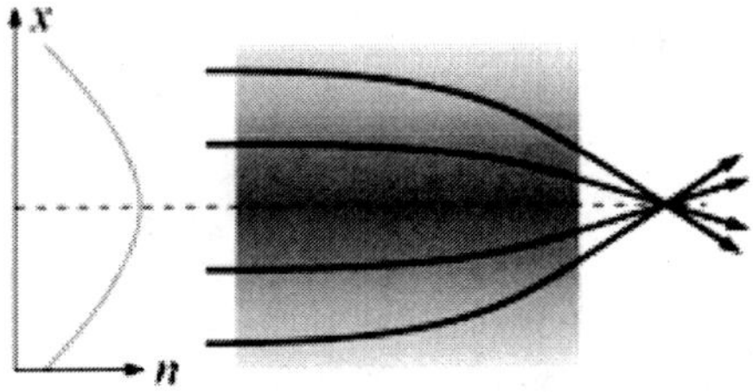

Figure : A gradient-index lens with a parabolic variation of refractive index (n) with radial distance (x). The lens focuses light in the same way as a conventional lens.

Refractive Index Measurement

Homogeneous Media

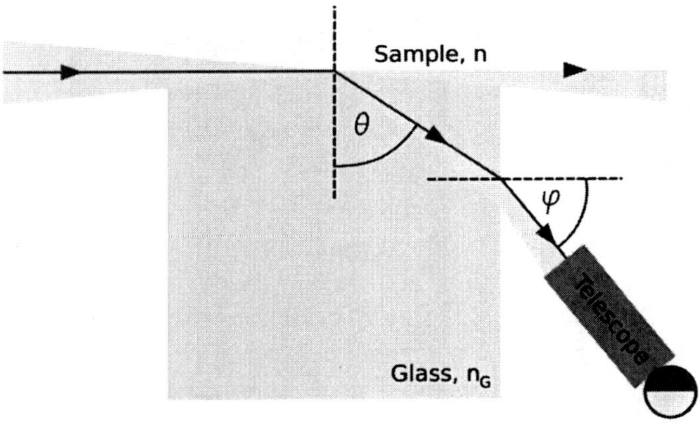

Figure: The principle of many refractometers.

The refractive index of liquids or solids can be measured with refractometres. They typically measure some angle of refraction or the critical angle for total internal reflection. The first laboratory refractometres sold commercially were developed by Ernst Abbe in the late 19th century. The same principles are still used today. In this instrument a thin layer of the liquid to be measured is placed between two prisms. Light is shone through the liquid at incidence angles all the way up to 90°, i.e. light rays parallel to the surface. The second prism should have an index of refraction higher than that of the liquid, so that light only enters the prism at angles smaller than the critical angle for total reflection. This angle can then be measured either by looking through a telescope, or with a digital photodetector placed in the focal plane of a lens. The refractive index n of the liquid can then be calculated from the maximum transmission angle θ as $n = n_G \sin \theta$, where n_G is the refractive index of the prism.

Figure : Photo of a handheld refractometre used to measure sugar content of fruits.

This type of devices are commonly used in chemical laboratories for identification of substances and for quality control. Handheld variants are used in agriculture by e.g.wine makers to determine sugar content in grape juice, and inline process refractometres are used in e.g. chemical and pharmaceutical industry for process control.

In gemology a different type of refractometre is used to measure index of refraction and birefringence of gemstones. The gem is placed on a high refractive index prism and illuminated from below. A high refractive index contact liquid is used to achieve optical contact between the gem and the prism. At small incidence angles most of the light will be transmitted into the gem, but at high angles total internal reflection will occur in the prism. The critical angle is normally measured by looking through a telescope.

4

Optics in Quantum

Colour Constancy

Colour constancy is an example of subjective constancy and a feature of the human colour perception system which ensures that the perceived colour of objects remains relatively constant under varying illumination conditions. A green apple for instance looks green to us at midday, when the main illumination is white sunlight, and also at sunset, when the main illumination is red. This helps us identify objects.

Physiological Basis

The physiological basis for colour constancy is thought to involve specialised neurons in the primary visual cortex that compute local ratios of cone activity, which is the same calculation that Land's retinex algorithm uses to achieve colour constancy. These specialised cells are called *double-opponent cells* because they compute both colour opponency and spatial opponency. Double-opponent cells were first described by Nigel Daw in the goldfishretina. There was considerable debate about the existence of these cells in the primate visual system; their existence was eventually proven using reverse-correlation receptive field mapping and special stimuli that selectively activate single cone classes at a time, so-called "cone-isolating" stimuli.

Colour constancy works only if the incident illumination contains a range of wavelengths. The different cone cells of the eye register different ranges of wavelengths of the light reflected by every object in the scene. From this information, the visual system attempts to determine the approximate composition of the illuminating light. This illumination is then *discounted* in order to obtain the object's "true colour" or reflectance: the wavelengths of light the object reflects. This reflectance then largely determines the perceived colour.

Figure: Colour constancy: The colours of a hot air balloon are recognised as being the same in sun and shade

Figure: Automatically making assumptions about reflectance and the type and source of light

Figure: Constancy makes square A appear darker than square B, when in fact they are both exactly the same shade of grey.

Retinex Theory

The effect was described in 1971 by Edwin H. Land, who formulated "retinex theory" to explain it. The word "retinex" is a portmanteau formed from "retina" and "cortex", suggesting that both the eye and the brain are involved in the processing. The effect can be experimentally demonstrated as follows. A display called a "Mondrian" (after Piet Mondrian whose paintings are similar) consisting of numerous coloured patches is shown to a person. The display is illuminated by three white lights, one projected through a red filter, one projected through a green filter, and one projected through a blue filter. The person is asked to adjust the intensity of the lights so that a particular patch in the display appears white. The experimenter then measures the intensities of red, green, and blue light reflected from this white-appearing patch. Then the experimenter asks the person to identify the colour of a neighbouring patch, which, for example, appears green. Then the experimenter adjusts the lights so that the intensities of red, blue, and green light reflected from the green patch are the same as were originally measured from the white patch. The person shows colour constancy in that the green patch continues to appear green, the white patch continues to appear white, and all the remaining patches continue to have their original colours.

Colour constancy is a desirable feature of computer vision, and many algorithms have been developed for this purpose. These include several retinex algorithms. These algorithms receive as input the red/green/blue values of each pixel of the image and attempt to estimate the reflectances of each point. One such algorithm operates as follows: the maximal red value r_{max} of all pixels is determined, and also the maximal green value g_{max} and the maximal blue value b_{max}. Assuming that the scene contains objects which reflect all red light, and (other) objects which reflect all green light and still others which reflect all blue light, one can then deduce that the illuminating light source is described by $(r_{max}, g_{max}, b_{max})$. For each pixel with values (r, g, b) its reflectance is estimated as $(r/r_{max}, g/g_{max}, b/b_{max})$. The original retinex algorithm proposed by Land and McCann uses a localised version of this principle.

Although retinex models are still widely used in computer vision, they have been shown not to accurately model human colour perception.

Colour Phi Phenomenon

The Colour Phi phenomenon is a perceptual illusion described by psychologists Paul Kolers and Michael von Grunau in which a disembodied perception of motion is produced by a succession of still images. The Colour Phi phenomenon is a more complex variation of the Phi phenomenon. Kolers and von Grunau originally investigated the phenomenon in response to a question posed by the philosopher Nelson Goodman, who asked what the effect of the colour change would have on the Phi phenomenon.

The classic colour phi phenomenon experiment involves a viewer or audience watching a screen, upon which the experimenter projects two images in succession. The first image depicts a blue dot at the top of the frame. The second image depicts a red dot on the bottom of the frame. The images may be shown quickly, in rapid succession, or each frame may be given several seconds of viewing time. Once both images have been projected, the experimenter asks the viewer or audience to describe what they saw. At certain combinations of spacing and timing of the two images, a viewer will report a sensation of motion in the space between two dots. The first spot will begin appear moving, and will then "change colour abruptly in the middle of its illusory path".

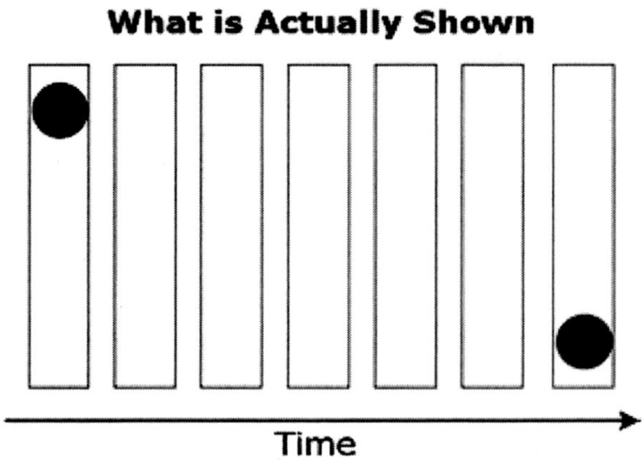

Figure: What is actually shown: First a blue dot is shown at the top of the screen, followed by a period of blank screen. Finally a red dot is shown at the bottom of the screen.

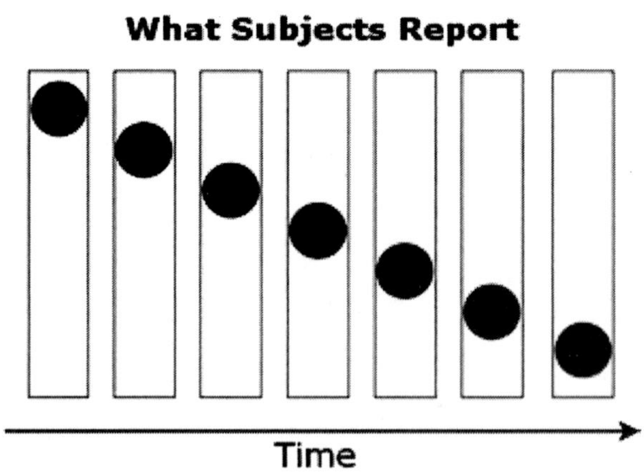

Figure: Subjects report seeing a dot that moves from the top to the bottom. The dot changes colour midway through its path.

Philosophical Significance

The existence of the colour phi phenomenon poses an interesting philosophical problem. When asked to describe their experience, subjects report seeing the abrupt colour change *before* the second dot is actually presented.

However, logically, it is impossible for a subject to actually experience the colour change before the second dot has been presented – how could the dot change colour before the second dot is shown and the subject finds out what the colour of the second dot is going to be? Philosopher Daniel Dennett utilises the colour phi phenomenon in his argument against a philosophy known as Cartesian materialism.

Contingent Perceptual Aftereffect

Contingent aftereffects are studied in human perception and can be defined as illusory percepts that are apparent on a test stimulus after exposure to an induction stimulus for an extended period.

Contingent aftereffects can be contrasted with simple aftereffects, the later requiring no test stimulus for the illusion/mis-perception to be apparent.

Contingent aftereffects have been studied in different perceptual domains. For instance, visual contingent aftereffects, auditory contingent aftereffects and haptic contingent aftereffects have all been discovered.

An example of a visual contingent aftereffect is the McCollough effect. The McCollough Effect is one of a family of contingent aftereffects related to the processing of colour and orientation. One can induce the aftereffect by exposure to a magenta and black vertical grating alternating with a green and black horizontal grating.

After a few minutes of induction (5 or more is best), followed by a break of a few minutes, black-and-white vertical and horizontal gratings will appear coloured.

The verticals will look green and horizontals pink in the example given. Therefore, the illusory colour apparent on the test fields is *contingent* on the orientation of the lines in that test field.

Furthermore, the orientation-colour contingencies present in the illusion are the *reverse* of those present in the adapting stimulus (i.e., the magenta-vertical and green-horizontal adaptation gratings produced illusory magenta on the horizontal test gratings and illusory green on the vertical test grating).

The illusion will reverse if one rotates one's head 90°. This is because the effect is retino-topic, that is, the effect is dependent on the orientation of the test lines on the retina. There are also colour-contingent motion aftereffects, and other varieties of these phenomena.

Convergence Micropsia

Convergence micropsia is a type of micropsia characterised by the reduction in apparent size of objects viewed when the eyes are more converged than they need to be for the distance of the object from the eyes.

It occurs mainly during stereoscopy and when viewing autostereograms (such as Magic Eye pictures). In these cases, the object is depicted by the two half images of a stereogram or by the contents of the autostereogram. Moving the stereogram or the autostereogram closer to the eyes increases convergence of the eyes and reduces the apparent size of the depicted object. If a correctly arranged stereogram or autostereogram is viewed with crossed eyes, then the depicted objects will appear smaller than if it is viewed with eyes diverged or parallel.

Convergence micropsia could also occur if one looked at the world through base-out prisms.

Explanation

Apparent size is influenced by the size of the retinal image of an object and by its apparent distance from the eyes. This is shown clearly by Emmert's law, in which the apparent size of an afterimage (e.g., the bright spot we see after looking at a camera flash) is influenced by where it is viewed.

An afterimage has a fixed size on the retina, resulting from adaptation of the rod cells and cone cells of the retina. When an afterimage is viewed on the far wall of a room it looks large; when it is viewed on one's hand, it looks small. It is consistent with the geometry of the world: an object with a particular visual angle (given by the size of the retinal image) must be large if it is far away and small if it is near.

Convergence micropsia implies that the state of convergence of the eyes contributes to determining the distance of an object from the eyes, that it acts as a depth cue. At six metres, to view an object without double vision the optic axes of the eyes are essentially parallel (no convergence). At closer distances, to view an object without double vision the optic axes need to approach each other at an angle (increasing convergence). Normally, the convergence angle gives accurate information about the distance of objects. Under the conditions that yield convergence micropsia, the overconverged vergence angle specifies a shorter distance than the actual distance of the object; this affects apparent size.

Cornsweet Illusion

Figure: Cornsweet illusion. Left part of the picture seems to be darker than the right one. In fact they have the same brightness.

Figure: The same image as above, but the edge in the middle is hidden. Left and right part of the image appear as the same colour now.

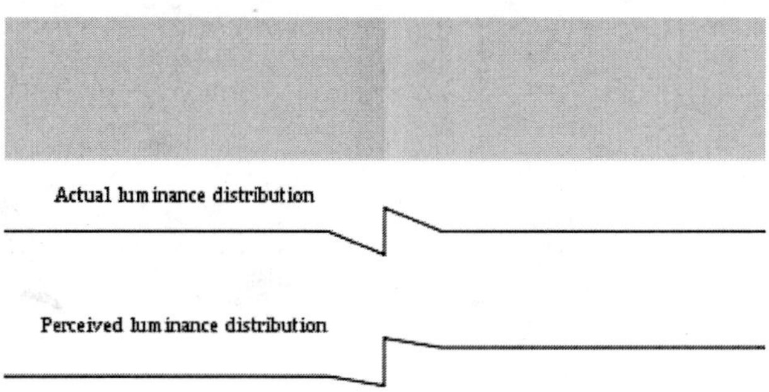

Figure: The actual distribution of luminance in the picture, and the typical perception of luminance.

The Cornsweet illusion, also known as *Craik–O'Brien–Cornsweet illusion* and *Craik–Cornsweet illusion*, is an optical illusion that was described in detail by Tom Cornsweet in the late 1960s.

Craik and O'Brien had made earlier observations in a similar vein. In the image at right, the entire region to the right of the "edge" in the middle looks slightly lighter than the area to the left of the edge, but in fact the brightness of both areas is exactly the same, as can be seen by blacking out the region containing the edge.

This phenomenon is similar to the phenomenon of simultaneous contrast and Mach bands, but differs from it in two important respects.

- In Mach bands, the effect is seen only on areas that are close to the intensity gradient. In Craik–O'Brien–Cornsweet illusion, a very small area (the central "edge") affects the perception of entire large areas, portions of which are distant from the edge.
- In the Cornsweet illusion, the region adjacent to the light part of the edge appears lighter, and the region adjacent to the dark part of the edge appears darker, just the opposite of the usual contrast effects.

The third image at right shows the usual explanation of the effect. A far more convincing and dramatic version of the effect can be seen in the article by Purves, Lotto, and Nundy, where it is presented within a quasi-realistic image of solid, illuminated objects.

These writers give a explanation of this and other illusions, in which the visual system and brain are posited to generate percepts on an empirical basis that is much like a reflex.

In their words, *...[perception] accords not with the features of the retinal stimulus or the properties of the underlying objects, but with what the same or similar stimuli have typically signified in the past.*

Delboeuf Illusion

The two black circles are exactly the same size; however, the one on the left seems larger.

The Delboeuf illusion is an optical illusion of relative size perception. In the best-known version of the illusion, two circles of identical size have been placed near to each other and one is surrounded by anannulus; the surrounded circle then appears larger than the non-surrounded circle if the annulus is close, while appearing smaller than the non-surrounded circle if the annulus is distant. Recent work suggests it is caused by the same visual processes that cause the Ebbinghaus illusion

Disappearing Model

Disappearing Model is a trompe-l'œil body painting by Joanne Gair that was part of episode 1.19 or 119 of Ripley*'s Believe It or Not!*, which was the highest rated episode. The number 119 represents the nineteenth episode of the first season of the show, which aired exclusively on the TBS superstation.

General Episode Details

The episode was hosted by Dean Cain, and it originally aired on either August 30 or September 6, 2000, depending upon the source. Both TV.com and IMDb.com claim the episode aired on the August date. TBS and *Business Wire* sources dated August 31, 2000 announce its premiere on the September date. Sony Pictures claims the episode aired on November 15, 2000, but does not clarify whether this was the premiere airing or arerun. The episode had a total of six features.

Gair's Role on the Episode

Descriptions of the episode describe Gair's notability for her ability to "...make people disappear by painting them right into a background or paint clothing on a human body that is virtually undistinguishable from actual fabric!" More detailed descriptions of the episode describe the origin of Gair's notability as deriving from her production of the *Vanity Fair* magazine cover entitled *Demi's Birthday Suit*, also known as "The Suit" in 1992. During the episode, Gair exhibits body painting that appears to be no-fabric clothing and creates another work that causes the illusion that a model disappears into the background of a flower print. The 2000 body painting in which the model is indistinguishable from the red, blue and yellow flowers is known as *Disappearing Model* and is considered to be Gair's most famous work. Gair's second book, *Body Painting: Masterpieces by Joanne Gair*, includes a photo of *Disappearing Model*. The section credits Tracy Bayne as the photographer of the Russian model Sasha. Gair notes that the show's producer, Gail Smerigan contacted her in 2000 about featuring her work in an episode. The intent was to capture a photo shoot specifically designed for the show with the intent of creating the illusion of causing the model to disappear.

Ebbinghaus Illusion

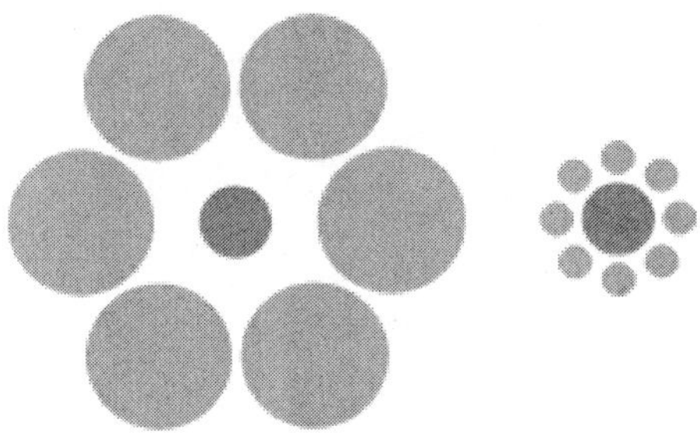

Figure: The two orange circles are exactly the same size; however, the one on the left seems smaller.

The Ebbinghaus illusion or Titchener circles is an optical illusion of relative size perception. In the best-known version of the illusion, two circles of identical size are placed near to each other and one is surrounded by large circles while the other is surrounded by small circles; the first central circle then appears smaller than the second central circle. It was named for its discoverer, the German psychologist Hermann Ebbinghaus (1850-1909) it was popularised in the English-speaking world by Titchener in a 1901 textbook of experimental psychology, hence its alternative name *"Titchener circles"*. Although commonly thought of as an illusion of *size,* recent work suggests that the critical factor in the illusion is the distance of the surrounding circles and the completeness of the annulus, making the illusion a variation of the Delboeuf illusion. If the surrounding circles are near to the central circle it appears larger, while if they are far away it appears smaller. Obviously, the size of the surrounding circles dictates how near they can be to the central circle, resulting in many studies confounding the two variables. The Ebbinghaus illusion has played a crucial role in the recent debate over the existence of separate pathways in the brain for perception and action. It has been argued that the Ebbinghaus illusion distorts *perception* of size, but when a subject is required to respond with an *action*, such as grasping, no size distortion occurs. However, recent work suggests that the original experiments were flawed. The original stimuli limited the possibility for error in the grasping action, therefore making the grasp response more accurate, and presented the large and small versions of the stimulus in isolation—which results in *no illusion* because there is no second central circle to act as a reference. Franz *et al.* conclude that both the action and perception systems are equally fooled by the Ebbinghaus illusion.

Other recent research suggests that an individual's receptivity to this illusion, as well as the Ponzo illusion, may be inversely correlated with the size of that individual's primary visual cortex.

Sometimes the name "Ehrenstein" is associated with one of the illusory contour. The ends of the dark segments produce the illusion of circles. The apparent figures have the same colour as the background, but appear brighter. A similar effect is obtained in the Kanizsa triangle.

Ehrenstein Illusion

The Ehrenstein illusion is an optical illusion studied by the German psychologist Walter Ehrensteinin which the sides of a square placed inside a pattern of concentric circles take an apparent curved shape (1).

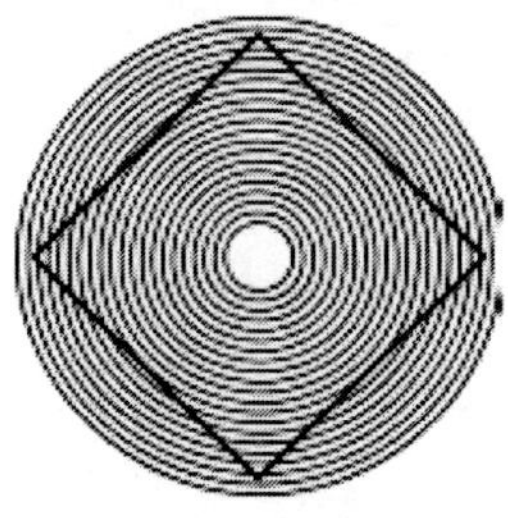

Figure: Ehrenstein illusion

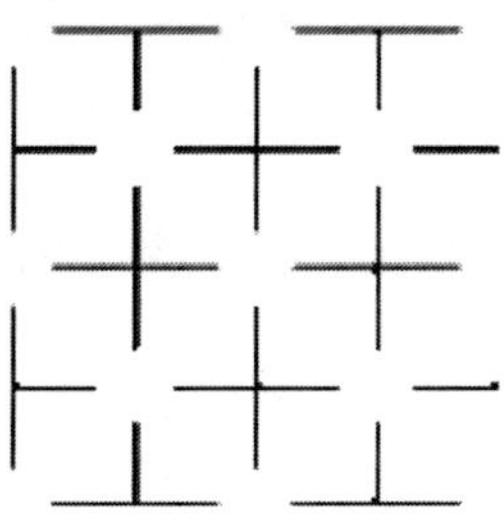

Figure: Illusion

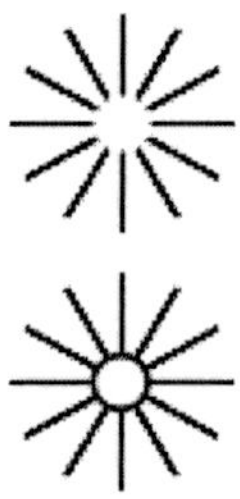

Figure: Illusion; adding a circle (bottom) destroys the illusion of a bright central disk.

Fechner Colour

The Fechner colour effect is an illusion of colour seen when looking at certain rapidly changing or moving black-and-white patterns. They are also called pattern induced flicker colours (PIFCs). Not everyone sees the same colours.

The effect is most commonly demonstrated with a device known as Benham's top. It can also be seen in stroboscopic lights when flashes are set at certain critical speeds. Rotating fan blades, particularly aluminium ones, can also demonstrate the effect; as the fan accelerates or decelerates, the colours appear, drift, change and disappear.

The stable running speed of the fan does not (normally) produce colours, suggesting that it is not an interference effect with the frequency of the illumination flicker.

The effect was noted by Gustav Fechner and Hermann von Helmholtz and propagated to English-speakers through Charles Benham's invention of his top. The perceptual mechanism of Fechner colour is not entirely understood.

When the disk is spun, arcs of pale colour are visible at different places on the disk. One possible reason people see colours may be that the colour receptors in the human eye respond at different rates to red, green, and blue.

Or, more specifically, that the latencies of the centre and the surrounding mechanisms differ for the different types of colour-specific ganglion cells.

The phenomenon originates from neural activity in the retina and spatial interactions in the primary visual cortex, which processes pattern recognition. (von Campenhausen & Schramme, 1995) Research indicates that the blue-yellow opponent process accounts for all the different PIFCs. (Schramme, 1992)

Figure–Ground (Perception)

Figure–ground organisation is probably best known by the faces–vase drawing that Edgar Rubin described. This drawing exemplifies one of the key aspects of figure–ground organisation, edge-assignment and its effect on shape perception. Notice in the faces/vase drawing below, the perceived shape depends critically on the direction in which the border (edge) between the black and white regions is assigned. If the two curvy edges between the black and white regions are assigned inward then the central white region is seen as a vase shape in front of a black background.

No faces are perceived in this case. On the other hand, if the edges are assigned outwards, then the two black profile faces are perceived on a white background and no vase shape is perceived. Studying figure-ground processes is a main issue in gestalt psychology.

Details

Figure–ground organisation is used to help artists and designers in composition of a 2D piece. In its basic sense, it refers to a cognitive ability to separate elements based upon contrast, that is, dark and light, black and white. Many times this definition is expanded from a simple perception based on contrast to include abstract (i.e. non-visual) concepts such as melody/harmony, subject/background and positive/negative space. The famous optical illusion image depicting both a vase and two profiles of a human face (the Rubin vase) is often used to illustrate the concept of figure ground: Depending on whether the white or black colour is seen as the figure (forefront) or the ground (background) the brain will interpret the picture as two different images, and it may be difficult (or even, according to gestalt psychological theory, impossible) to perceive both meaningful images simultaneously.

There is a tendency to switch rapidly between both 'readings' of the image. The Flag of Canada has also been cited as an example of figure–ground reversal, in which the background edges of the maple leaf can also be seen as two facesarguing.

Figure–ground reversal may also be used an intentional visual design technique in which an existing image's foreground and background colours are purposely swapped to create new images.

Types of Figure–ground Problems

There are three types of figure–ground problems:

1. The figure and the ground compete.

2. The figure should be the ground and the ground should be the figure.
3. The figure and ground create an optical illusion.

Filling-in

Figure: When steadily fixating the central dot for many seconds, the peripheral annulus will fade and will be replaced by the colour or texture of the background.

In vision, filling-in phenomena are those responsible for the completion of missing information across the physiological blind spot, and across natural and artificials cotomata. There is also evidence for similar mechanisms of completion in normal visual analysis. Classical demonstrations of perceptual filling-in involve images stabilised on the retina either by means of special lenses, or under certain conditions of steady fixation.

If a stimulus is entirely stabilised, its colour and lightness fade until they are no longer seen and the area fills in with the colour and lightness of the surrounding region. A famous example of fading under steady fixation is Troxler's fading. When steadily fixating on the central dot for many seconds, the peripheral annulus will fade and will be replaced by the colour or texture of the background. Since the adapted region is actively filled-in with background colour or texture, the phenomenon cannot be fully explained by local processes such as adaptation.

There is general agreement that edges play a central role in determining the apparent colour and lightness of surfaces through similar filling-in mechanisms. However, the way in which their influence is performed is still unclear. Two different theories have been put forward to explain the filling-in completion phenomenon.

One theory % addressed as the 'Isomorphic filling-in theory' according to the definition of Von der Heydt, Friedman *et al.* (2003) % postulates that perception is based on an image representation held in a two dimensional array of neurons, typically arranged retinotopically, in which colour signals spread in all directions except across borders formed by contour activity. The process is thought to be analogous to physical diffusion, with contours acting as diffusion barriers for the colour and brightness signals. An alternative hypothesis is that image information is transformed at the cortical level into an oriented feature representation. Form and colour would be derived at a subsequent stage, not as the result of an isomorphic filling-in process, but as an attribute of an object or proto-object. This theory is called the symbolic filling-in theory.

According to the isomorphic filling-in theory, colour is represented by the activity of cells whose receptive fields point at the surface, but it is assumed that these cells receive additional activation through horizontal connections that keeps their activity level high despite mechanisms of lateral inhibition tending to suppress surface activity and despite the transient nature of the afferent signals. The lateral activation comes from receptive fields at contrast borders.

These signals are strong because receptive fields are exposed to contrast, and reliable because the border produces continuous light modulation even during fixation, due to small residual eye movements. In the alternative symbolic hypothesis, there is no spreading of activity, but all the information would be carried by the relevant features, that would be tagged with information on contrast polarity, colour and lightness of the surfaces they enclose. Despite the many attempts to verify the two different models by psychophysical and physiological experiments, the mechanisms of colour and lightness filling-in are still debated.

Isomorphic Filling-in

There are at least three different kinds of experiments whose results support the idea that a real spreading of neural activity in early visual areas is the basis for filling-in of visual information.

Recordings from Cells of the Blind Spot Representation in Monkey Striate Cortex

Komatsu and colleagues recorded activity of cells of the blind spot representation in monkey striate cortex (area V1) and found some cells, in layers 4-6, that responded to large stimuli covering the blind spot (the condition under which filling- in is perceived), but not to small stimuli near the blind spot. A neuronal circuitry seems to exist that elaborates and transmits colour and brightness information through the blind region.

Though intriguing, these results cannot be easily generalised to similar phenomena, such as the filling-in of illusory contours or the filling-in through artificial scotomata or adapted edges (such as in the Troxler's effect). All these phenomena are indeed similar, and probably rely on similar neural circuitries but they are not identical. For instance, an obvious difference between filling-in across the blind spot and filling-in of occluded edges is that filling-in across the blind spot is modal (i.e. you literally see the filled-in section), while filling-in across occluders is a modal. Filling-in across the blind spot was found to be different also from filling-in across cortical scotomata in two patients examined by Ramachandran (Ramachandran 1992; Ramachandran, Gregory *et al.* 1993). In these subjects, some features filled in the scotoma faster than others, and in some circumstances filling-in took some seconds before it was completed (while filling-in across the blind spot is immediate). Together these data suggested that mechanisms for the filling-in of colours, motion and texture can be dissociated and may correspond to processes in higher-order areas that are specialised for these attributes.

Delayed Masking Psychophysical Experiments

Striking evidence implying a spreading of neural activity like the one postulated by isomorphic filling-in theory is given by experiments of backward masking after brief presentations of uniform surfaces or textures. The working hypothesis of these experiments is that if a response initially biased toward the boundaries fills-in to represent the interiors of uniform surfaces, it may be possible to interfere with the filling-in process and leave the percept at an incomplete stage.

Paradiso and Nakayama (1991) performed an experiment to verify this hypothesis. They presented a large disk of uniform brightness on a black background. The stimulus was briefly flashed and, after a variable stimulus offset a synchrony, a masking stimulus was presented.

The mask consisted of a circle on a black background with the masking contours positioned within the boundaries of the large uniform disk. This experiment is grounded on the assumption that filling-in consists of a spreading of neural activity from the boundaries of luminance and through the surfaces, that is stopped when another luminance-contrast border is reached (this is proposed by many models of brightness perception.

Subjects were asked to match the brightness at the centre of the disk with a palette of grey scales. When the delay between target and mask presentation was long enough, the mask had no effect on the apparent brightness of the stimulus, but for stimulus offset asynchronies of 50-100ms, the surface of the disk inside the masking annulus appeared unfilled. Moreover, the minimum target-mask delay at which the masking was effective increased with target size, suggesting that there would be a spreading phenomenon and that the farther the features delimiting a region, the more time is necessary for the filling-in to be completed. These results are supported also by further experiments on temporal limits of brightness induction in simultaneous contrast (De Valois, Webster *et al.* 1986; Rossi and Paradiso 1996; Rossi, Rittenhouse *et al.* 1996), as well as by a similar experiment performed by Motoyoshi (1999) on filling-in of texture.

Recordings from Striate Cortex of Cat

An isomorphic filling-in theory calls for the existence of surface responsive neurons in early retinotopic visual areas. The activity of such neurons would be raised by elements capable of responding to the luminance of the surface also in the absence of edges; and would be strongly modulated by spreading of activity from the luminance borders enclosing the surface.

Electro-physiological recordings in retinal ganglion cells, LGN and primary visual cortex showed that neurons of these areas responded to luminance modulation within the receptive field even in the absence of contrast borders.

In a second condition, a uniform grey patch was placed on the receptive field (extending 3- 5 degrees beyond the receptive field boundary on either side), and two flanking patches modulated sinusoidally in time from dark to light. With such stimuli, the brightness of the central patch appears to modulate, despite the absence of luminance change. In this condition, cat retinal ganglion cells and lateral geniculate nucleus cells, having their receptive fields centred in the uniform grey patch, did not respond; on the other hand, primary visual cortex neurons were modulated by luminance changes far outside their receptive fields. Together, these results suggest that neurons in the retina and LGN are responsive to luminance modulation, but their response does not correlate with perceived brightness. On the other hand, striate neurons responded to stimulus conditions producing changes in brightness in the area corresponding to the receptive field. The behaviour of primary visual cortex neurons seems to be in agreement with the one hypothesized by an isomorphic filling-in theory in that they both respond to luminance of the surfaces also in the absence of borders, and their activity is modulated by that of edges far outside the receptive field. Moreover, when the temporal frequency of luminance modulation in the surrounding patches exceeded a threshold value, the induced response disappeared, suggesting that it was the result of a spreading of activity, taking a finite time to happen, likely explainable in the context of isomorphic filling-in.

Symbolic Filling-in

"Perceptual filling-in", in its simplest definition, is simply the filling-in of information that is not directly given to the sensory input. The missing information is inferred or extrapolated from visual data acquired in a different part of the visual field. Examples of filling-in phenomena include lightness assignment to surfaces from information of contrast across the edges and completion of features and textures across the blind spot, based on the features and textures that are detected in the visible part of the image. In this definition, it is clear that a filling-in process involves a rearrangement of visual information, in which activity in one region of the visual field (i.e. edges) is assigned to other regions (surfaces). In any event, the total amount of information available is not increased, being determined by the retinal input, and any rearrangement of information is useful only if it brings the information contained in the image into a form that is more easily analysed by our brain.

Dennet and Kinsbourne (Dennet 1992; Dennet and Kinsbourne 1992) opposed to the idea that an active filling-in process would take place in our brain on philosophical grounds. They argued that such an idea would be the result of the false belief that in our brain there is a spectator, a sort of homunculus similar to ourselves, needing a filled-in image representation. From a scientific viewpoint, Dennet's homunculus may correspond to higher-order scene representation or decision-making mechanisms. The question is whether or not such mechanisms need a filled in, gap free representation of the image to function optimally (Ramachandran 2003). The symbolic filling-in theory postulates that such a "homunculus" need not exist, and that image information is transformed at the cortical level into an oriented feature representation. Surface form and colour are not coded at this stage, but would be derived only at a symbolic level of representation, as attributes of objects or proto-objects.

Electrophysiology Recordings from Monkey Primary Visual Cortex

When experiencing perceptual filling-in, the colour of an adapted surface is gradually replaced by the colour or texture of outside the surface. Friedman *et al.* (Friedman 1998; Friedman, Zhou *et al.* 1999) performed an experiment aimed at determining if surface activity of cells in monkey primary visual cortex changed in accordance with perceptual change or simply followed the modulation of the colour presented to the retina.

Stimuli consisted of a disk-ring configuration similar to that illustrating the Troxler effect, but where the inner and outer part of the annulus have two physically different colours. After a few seconds of (peripheral) fixation, the disk tends to disappear, whereas the outer contour of the ring is perceived much longer, and the area of the disk is filled-in with the colour of the ring (Krauskopf 1967).

These stimuli where intermixed with control stimuli, in which the physical colour of the disk was gradually changed to that of the ring. The animals were instructed to signal a colour change, and their responses to control stimuli and to test stimuli were compared in order to determine if monkeys perceive colour filling-in under steady fixation like humans.

The authors recorded the activity of surface- and edge-cells (cells whose receptive fields pointed either to the filled-in surface or to the border between the disk and the ring) in the visual cortices V1 and V2 while the monkey was performing the filling-in task. The activity of surface-cells correlated with the physical stimulus change in both areas V1 and V2, but not with the perceived colour change induced by filling-in. The activity of edge-cells followed the stimulus contrast when the disk colour changed physically; when the colours were constant, the edge signals also decayed, but more slowly. Together, these data are incompatible with the isomorphic filling-in theory, which assumes that colour signals spread from the borders into uniform regions.

fMRI Experiments in Human Subjects

The neuronal activity in different brain areas can be recorded in humans through non-invasive techniques, like fMRI (functional magnetic resonance imaging). Perna *et al.* (2005) used fMRI to investigate the neuronal mechanisms responsible for the Craik-O'Brien-Cornsweet illusion.

These authors recorded the activity in different brain areas when observers were presented with a Cornsweet visual stimulus, and compared the activities with those elicited by a similar image, which however did not elicit any brightness filling-in.

Contrary to the predictions of isomorphic filling-in, these authors found an identical response to the stimulus that induced filling-in and to the control stimulus in early visual cortex. Recently, Cornelissen *et al.* (2006) performed a similar experiment involving the simultaneous contrast illusion. These authors presented observers in an fMRI scan with simultaneous contrast stimuli composed of a central circle of uniform luminance and a peripheral region whose luminance was modulated in time (and also tested other conditions where the modulated and the constant regions were inverted). The brain activity was recorded in primary visual cortex in a retinotopic position corresponding to the perceptually filled-in region. Also in this condition, no activity was found at this level in response to the filled-in signal.

Flash Lag Illusion

The flash lag illusion or flash-lag effect is a visual illusion wherein a flash and a moving object that appear in the same location are perceived to be displaced from one another (MacKay, 1958; Nijhawan, 1994). Several explanations for this simple illusion have been explored in the neuroscience literature.

Motion Extrapolation

The first proposed explanation for the flash-lag effect is that the visual system is predictive, accounting for neural delays by extrapolating the trajectory of a moving stimulus into the future (Nijhawan, 1994; Khurana and Nijhawan, 1995). In other words, when light from a moving object hits the retina, a certain amount of time is required before the object is perceived. In that time, the object has moved to a new location in the world. The motion extrapolation hypothesis asserts that the visual system will take care of such delays by extrapolating the position of moving objects forward in time.

Latency Difference

A second proposed explanation is that the visual system processes moving objects more quickly than flashed objects. This latency-difference hypothesis asserts that by the time the flashed object is processed, the moving object has already moved to a new position (Baldo and Klein, 1995; Whitney & Murakami, 1998; Purushothaman et al., 1998). The latency-difference proposal tacitly rests on the assumption that awareness (what the subject reports) is an on-line phenomenon, coming about as soon as a stimulus reaches its "perceptual end-point" (Zeki & Bartels, 1998).

Motion Integration and Postdiction

Eagleman & Sejnowski (2000abc) proposed a third alternative: visual awareness is neither predictive nor on-line, but is instead *postdictive*, such that the percept attributed to the time of the flash is a function of events that happen in the ~80 msec following the flash. This postdictive framework is consistent with findings in other fields, such as backward masking in visual psychophysics (Bachmann, 1994), or the Colour Phi phenomenon (Kolers & von Grunau, 1976). In backward masking, a stimulus followed in rapid succession by a second stimulus can block or modify the perception of the first one.

In the colour phi phenomenon, 2 coloured dots presented sequentially within a small time and distance will appear to have changed colour in the middle of their apparent trajectory. Since the viewer cannot know what the colour of the second dot will be until having seen the second dot, the only explanation is that the conscious percept attributed to the 'trajectory' of the dots is formed after the second dot has 'arrived' at its destination.

Eagleman & Sejnowski found that the perception attributed to the time of the flash depends on events in the next ~80 msec after the flash. In this way, they drew a correspondence between the flash-lag effect and the Frφhlich effect (Frφhlich, 1923), wherein the first position of a moving object entering a window is misperceived.

Forced Perspective

Forced perspective is a technique that employs optical illusion to make an object appear farther away, closer, larger or smaller than it actually is. It is used primarily in photography, filmmaking and architecture. It manipulates human visual perception through the use of scaled objects and the correlation between them and the vantage point of the spectator or camera.

Forced Perspective in Filmmaking

Examples of forced perspective:

- A scene in an action/adventure movie in which dinosaurs are threatening the heroes. By placing a miniature model of a dinosaur close to the camera, the dinosaur may look monstrously tall to the viewer, even though it is just closer to the camera.

Movies (especially B-movies) in the 1950s and 1960s produced on limited budgets sometimes feature forced perspective shots which are completed without the proper knowledge of the physics of light used in cinematography, so foreground models can appear blurred or incorrectly exposed. Forced perspective can be made more believable when environmental conditions obscure the difference in perspective. For example, the final scene of the famous movie *Casablanca* takes place at an airport in the middle of a storm, although the entire scene was shot in a studio.

This was accomplished by using a painted backdrop of an aircraft, which was "serviced" by dwarfs standing next to the backdrop. A downpour (created in-studio) draws much of the viewer's attention away from the backdrop and extras, making the simulated perspective less noticeable.

Role of Light

Early instances of forced perspective used in low-budget motion pictures showed objects that were clearly different from their surroundings: often blurred or at a different light level.

The principal cause of this was geometric. Light from a point source travels in a spherical wave, decreasing in intensity (or illuminance) as the inverse square of the distance travelled. This means that a light source must be four times as bright to produce the same illuminance at an object twice as far away. Thus to create the illusion of a distant object being at the same distance as a near object and scaled accordingly, much more light is required.

Opening the camera's iris lets more light into the camera, allowing both near and far objects to be seen at a more similar light level, but this has the secondary effect of decreasing depth of field. This makes either the near or the far objects appear blurry. By increasing the volume of light hitting the distant objects, the iris opening can be restricted and depth of field is increased, thus portraying both near and far objects as in focus, and if well scaled, existing in a similar lateral plane.

Since miniature models would need to be subjected to far greater lighting than the main focus of the camera, the area of action, it is important to ensure that these can withstand the significant amount of heat generated by the incandescent light sources typically used in film and TV production.

Nodal Point: Forced Perspective in Motion

Peter Jackson's film adaptations of *The Lord of the Rings* make extended use of forced perspective. Characters apparently standing next to each other would be displaced by several feet in depth from the camera. This, in a still shot, makes some characters appear much smaller (for the dwarves and Hobbits) in relation to others. A new technique developed for *The Lord of the Rings: The Fellowship of the Ring* was an enhancement of this principle which could be used in moving shots. Portions of sets were mounted on movable platforms which would move precisely according to the movement of the camera, so that the optical illusion would be preserved at all times for the duration of the shot.

The same techniques were used in the Harry Potter movies to make the character Hagrid look like a giant. Props around Harry and his friends are of normal size, while seemingly identical props placed around Hagrid are in fact smaller.

The techniques developed centre around a nodal point axis, so the camera's panning axis is at the point between the lens and aperture ring where the light travelling through the camera meets its axis. By comparison, the normal panning axis would be at the point at which light would strike the film (orimage sensor in a digital camera).

The position of this nodal point can be different for every lens. However, on wide angle lenses it is often found between the midpoint of the lens and the aperture ring.

"Comedic" Effects

As with many film genre and effects, forced perspective can be used to visual-comedy effect. Typically, an object or character is portrayed in a scene, its size defined by its surroundings. A character then interacts with the object or character, in the process showing that the viewer has been fooled and there is forced perspective in use.

The 1930 Laurel and Hardy movie *Brats* used forced perspective to depict Stan and Olliesimultaneously as adults and as their own sons. An example used for comic effect can be found in the slapstick comedy *Top Secret!* in a scene which appears to begin as a close up of a ringing phone with the characters in the distance. However when the character walks up to the phone (towards the camera) and picks it up it becomes apparent that the phone is extremely oversized instead of being close to the camera. Another scene in the same movie begins with a closeup of a wristwatch. The next cut shows that the character actually has a gargantuan wristwatch.

The same technique is also used in the Dennis Waterman sketch in the British BBC sketch show Little *Britain*. In the television version, oversized props are used to make the caricatured Waterman look just three feet tall or less.

In *The History of the World, Part I*, while escaping the French peasants, Mel Brooks' character, Jacques, who is doubling for King Louis, runs down a hall of the palace, which turns into a ramp, showing the smaller forced perspective door at the end. As he backs down into the normal part of the room, he mutters, "Who designed this place?"

One of the recurring *The Kids in the Hall* sketches featured Mr. Tyzik, "The Headcrusher", who used forced perspective (from his own point of view) to "crush" other people's heads between his fingers.

In the making of Season 5 of *Red vs. Blue*, the creators used forced perspective to make the character of Tucker's baby look small. In fact in the game, the alien character used as the baby is the same height as other characters.

Forced Perspective in Architecture

In architecture, a structure can be made to seem larger, taller, farther away or otherwise by adjusting the scale of objects in relation to the spectator, increasing or decreasing perceived depth.

For example, when forced perspective is used to make an object appear farther away, the following method can be used: By constantly decreasing the scale of objects from expectancy and convention toward the farthest point from the spectator, an illusion is created that the scale of said objects is decreasing due to their distant location. In contrast, the opposite technique was sometimes used in classical garden designs and other "follies" to shorten the perceived distances of points of interest along a path.

The Statue of Liberty is built with a slight forced perspective so that it appears more correctly proportioned when viewed from its base. When the statue was designed in the late 19th century (before easy air flight), there were few other angles from which to view the statue. This caused a difficulty for special effects technicians working on the movie Ghostbusters II, who had to back off on the amount of forced perspective used when replicating the statue for the movie so that their model (which was photographed head-on) would not look top-heavy. This effect can also be seen in Michelangelo's statue of David.

Forced perspective is extensively employed at theme parks and other such architecture as found in Disneyland and Las Vegas, often to make structures seem larger than they are in reality where physically larger structures would not be feasible or desirable or to provide an optical illusion for entertainment value.

Fraser Spiral Illusion

Figure: Fraser spiral illusion

The Fraser spiral illusion is an optical illusion that was first described by the British psychologist James Fraser in 1908.

The illusion is also known as the *false spiral*, or by its original name, *the twisted cord illusion*. The overlapping black arc segments appear to form a spiral; however, the arcs are a series of concentric circles.

The visual distortion is produced by combining a regular line pattern (the circles) with misaligned parts (the differently coloured strands). Zφllner's illusion and the café wall illusion are based on a similar principle, like many other visual effects, in which a sequence of tilted elements causes the eye to perceive phantom twists and deviations. The illusion is augmented by the spiral components in the chequered background.

Auger Electron Spectroscopy

Auger electron spectroscopy (AES; pronounced [o'e] in French) is a common analytical technique used specifically in the study of surfaces and, more generally, in the area of materials science. Underlying the spectroscopic technique is the Auger effect, as it has come to be called, which is based on the analysis of energetic electrons emitted from an excited atom after a series of internal relaxation events. The Auger effect was discovered independently by both Lise Meitner and Pierre Auger in the 1920s.

Though the discovery was made by Meitner and initially reported in the journal Zeitschrift *für Physik* in 1922, Auger is credited with the discovery in most of the scientific community.

Until the early 1950s Auger transitions were considered nuisance effects by spectroscopists, not containing much relevant material information, but studied so as to explain anomalies in x-ray spectroscopy data. Since 1953 however, AES has become a practical and straightforward characterisation technique for probing chemical and compositional surface environments and has found applications in metallurgy, gas-phase chemistry, and throughout the microelectronics industry.

Electron transitions and the Auger Effect

The Auger effect is an electronic process at the heart of AES resulting from the inter- and intrastate transitions of electrons in an excited atom. When an atom is probed by an external mechanism, such as a photon or a beam of electrons with energies in the range of 2 keV to 50 keV, a core state electron can be removed leaving behind a hole. As this is an unstable state, the core hole can be filled by an outer shell electron, whereby the electron moving to the lower energy level loses an amount of energy equal to the difference in orbital energies. The transition energy can be coupled to a second outer shell electron which will be emitted from the atom if the transferred energy is greater than the orbital binding energy. An emitted electron will have a kinetic energy of:

$$E_{kin} = E_{\text{Core State}} - E_B - E_C'$$

where $E_{\text{Core State}}$, E_B, E_C' are respectively the core level, first outer shell, and second outer shell electron energies, measured from the vacuum level. The apostrophe (tic) denotes a slight modification to the binding energy of the outer shell electrons due to the ionized nature of the atom; often however, this energy modification is ignored in order to ease calculations. Since orbital energies are unique to an atom of a specific element, analysis of the ejected electrons can yield information about the chemical composition of a surface.

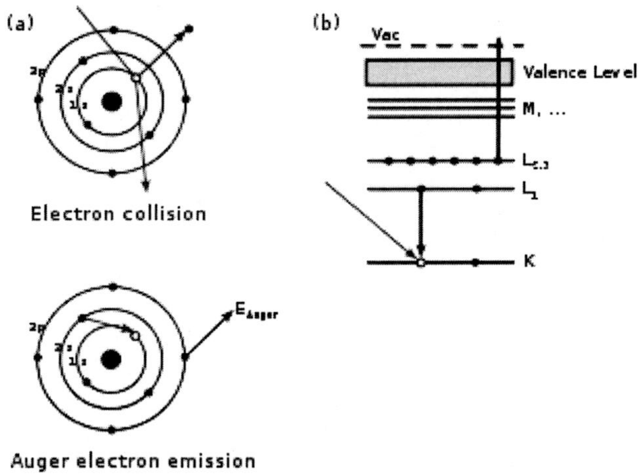

Figure: Two views of the Auger process. (a) illustrates sequentially the steps involved in Auger deexcitation. An incident electron creates a core hole in the 1s level. An electron from the 2s level fills in the 1s hole and the transition energy is imparted to a 2p electron which is emitted. The final atomic state thus has two holes, one in the 2s orbital and the other in the 2p orbital. (b) illustrates the same process using spectroscopic notation, $KL_1L_{2,3}$.

The types of state-to-state transitions available to electrons during an Auger event are dependent on several factors, ranging from initial excitation energy to relative interaction rates, yet are often dominated by a few characteristic transitions.

Because of the interaction between an electron's spin and orbital angular momentum (spin-orbit coupling) and the concomitant energy level splitting for various shells in an atom, there are a variety of transition pathways for filling a core hole. Energy levels are labelled using a number of different schemes such as the j-j coupling method for heavy elements ($Z \geq 75$), the Russell-Saunders L-S method for lighter elements ($Z < 20$), and a combination of both for intermediate elements.

The j-j coupling method, which is historically linked to X-ray notation, is almost always used to denote Auger transitions. Thus for a $KL_1L_{2,3}$ transition, Krepresents the core level hole, L_1 the relaxing electron's initial state, and $L_{2,3}$ the emitted electron's initial energy state. As figure illustrates this transition with the corresponding spectroscopic notation. The energy level of the core hole will often determine which transition types will be favoured. For single energy levels, i.e. K, transitions can occur from the L levels, giving rise to strong KLL type peaks in an Auger spectrum. Higher level transitions can also occur, but are less probable. For multi-level shells, transitions are available from higher energy orbitals (different n, $-\ell$ quantum numbers) or energy levels within the same shell (same n, different $-\ell$ number). The result are transitions of the type LMM and KLL along with faster Coster–Kronig transitions such as LLM. while Coster–Kronig transitions are faster, they are also less energetic and thus harder to locate on an Auger spectrum. As the atomic number Z increases, so too does the number of potential Auger transitions. Fortunately, the strongest electron-electron interactions are between levels which are close together, giving rise to characteristic peaks in an Auger spectrum. KLL and LMM peaks are some of the most commonly identified transitions during surface analysis. Finally, valence band electrons can also fill core holes or be emitted during KVV-type transitions. Several models, both phenomenological and analytical, have been developed to describe the energetics of Auger transitions. One of the most tractable descriptions, put forth by Jenkins and Chung, estimates the energy of Auger transition ABC as:

$$E_{ABC} = E_A(Z) - 0.5[E_B(Z) + E_B(Z + 1)] - 0.5[E_C(Z) + E_C(Z + 1)]$$

$E_i(Z)$ are the binding energies of the ith level in element of atomic number Z and $E_i(Z + 1)$ are the energies of the same levels in the next element up in the periodic table. While useful in practice, a more rigorous model accounting for effects such as screening and relaxation probabilities between energy levels gives the Auger energy as:

$$E_{ABC} = E_A - E_B - E_C - F(BC{:}x) + R_{xin} + R_{xex}$$

where $F(BC{:}x)$ is the energy of interaction between the B and C level holes in a final atomic state xa nd the R's represent intra- and extra-atomic transition energies accounting for electronic screening. Auger electron energies can be calculated based on measured values of the various E_i and compared to peaks in the secondary electron spectrum in order to identify chemical species. This technique has been used to compile several reference databases used for analysis in current AES setups.

Experimental Setup and Quantification

Instrumentation

Surface sensitivity in AES arises from the fact that emitted electrons usually have energies ranging from 50 eV to 3 keV and at these values, electrons have a short mean free path in a solid.

The escape depth of electrons is therefore localised to within a few nanometres of the target surface, giving AES an extreme sensitivity to surface species.

Because of the low energy of Auger electrons, most AES setups are run under ultra-high vacuum(UHV) conditions. Such measures prevent electron scattering off of residual gas atoms as well as the formation of a thin "gas (adsorbate) layer" on the surface of the specimen which degrades analytical performance.

A typical AES setup is shown schematically. In this configuration, focused electrons are incident on a sample and emitted electrons are deflected into a cylindrical mirror analyser (CMA).

In the detection unit, Auger electrons are multiplied and the signal sent to data processing electronics. Collected Auger electrons are plotted as a function of energy against the broad secondary electron background spectrum.

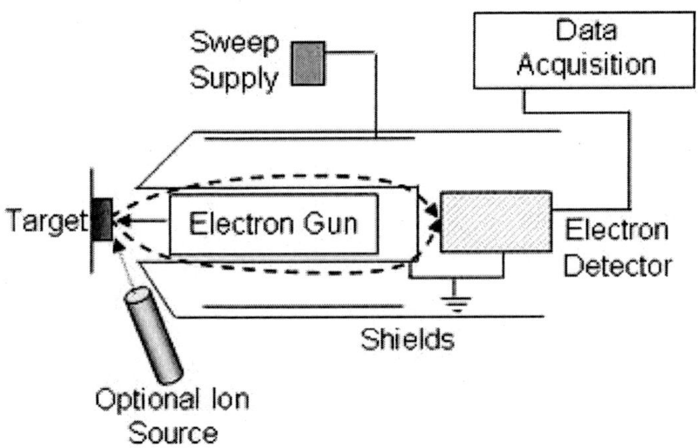

Figure : AES experimental setup using a cylindrical mirror analyser (CMA). An electron beam is focused onto a specimen and emitted electrons are deflected around the electron gun and pass through an aperture towards the back of the CMA. These electrons are then directed into an electron multiplier for analysis. Varying voltage at the sweep supply allows derivative mode plotting of the Auger data. An optional ion gun can be integrated for depth profiling experiments.

Since the intensity of the Auger peaks may be small compared to the noise level of the background, AES is often run in a derivative mode which serves to highlight the peaks by modulating the electron collection current via a small applied AC voltage. Since this $\Delta V = k\sin(\omega t)$, the collection current becomes $I(V + k\sin(\omega t))$. Taylor expanding gives:

$$I(V + k\sin(\omega t)) \approx I_0 + I'(V + k\sin(\omega t)) + O(I'')$$

Using the setup, detecting the signal at frequency ω will give a value for I' or $\frac{dN}{dE}$.Plotting in derivative mode also emphasizes Auger fine structure which appear as small secondary peaks surrounding the primary Auger peak. These secondary peaks, not to be confused with high energy satellites which are discussed later, arise from the presence of the same element in multiple different chemical states on a surface (i.e. Adsorbate layers) or from relaxation transitions involving valence band electrons of the substrate. It illustrates a derivative spectrum from a copper nitride film clearly showing the Auger peaks. The peak in derivative mode is not the true Auger peak, but rather the point of maximum slope of *N(E)*, but this concern is usually ignored.

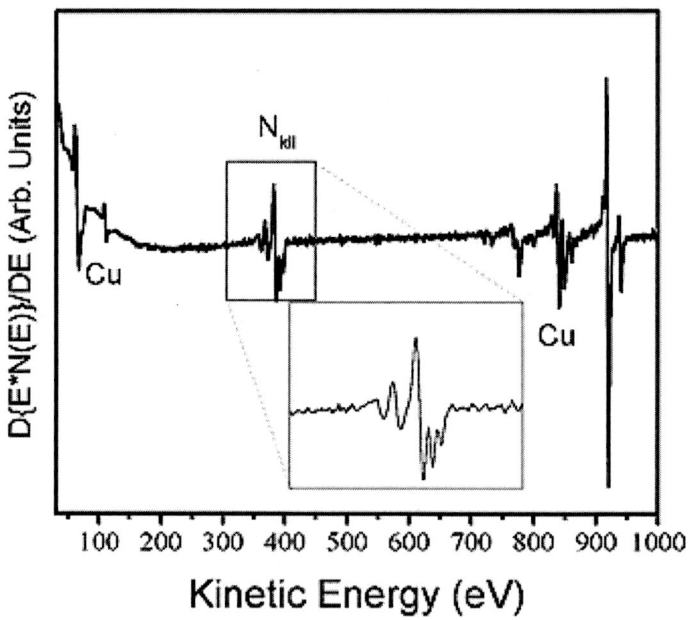

Figure : Auger spectrum of a copper nitride film in derivative mode plotted as a function of energy. Different peaks for Cu and N are apparent with the N KLL transition highlighted.

Quantitative Analysis

Semi-quantitative compositional and element analysis of a sample using AES is dependent on measuring the yield of Auger electrons during a probing event. Electron yield, in turn, depends on several critical parameters such as electron-impact cross-section and fluorescence yield. Since the Auger effect is not the only mechanism available for atomic relaxation, there is a competition between radiative and non-radiative decay processes to be the primary de-excitation pathway. The total transition rate, ω, is a sum of the non-radiative (Auger) and radiative (photon emission) processes. The Auger yield, ω_A, is thus related to the fluorescence(x-ray) yield, ω_X, by the relation,

$$\omega_A = 1 - \omega_X = 1 - \frac{W_X}{W_X + W_A}$$

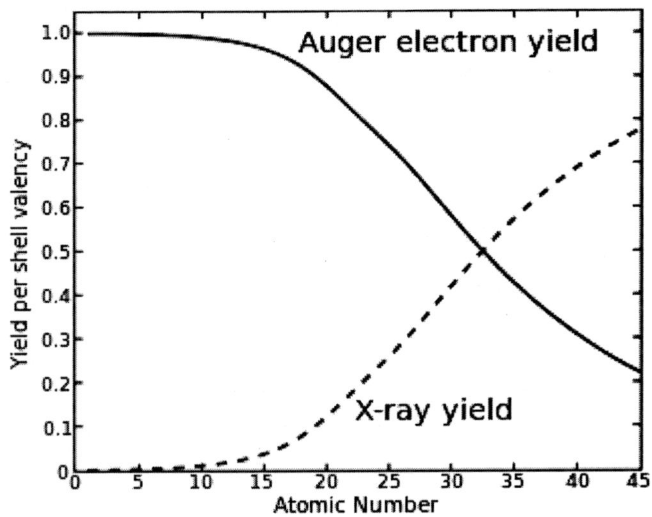

Figure : Fluorescence and Auger electron yields as a function of atomic number for K shell vacancies. Auger transitions (red curve) are more probable for lighter elements, while X-ray yield (dotted blue curve) becomes dominant at higher atomic numbers. Similar plots can be obtained for L and M shell transitions. Coster– Kronig (i.e. intra-shell) transitions are ignored in this analysis.

Where W_X is the X-ray transition probability and W_A is the Auger transition probability. Attempts to relate the fluorescence and Auger yields to atomic number have resulted in plots. A clear transition from electron to photon emission is evident in this chart for increasing atomic number. For heavier elements, x-ray yield becomes greater than Auger yield, indicating an increased difficulty in measuring the Auger peaks for large Z-values. Conversely, AES is sensitive to the lighter elements, and unlike X-ray fluorescence, Auger peaks can be detected for elements as light aslithium ($Z = 3$). Lithium represents the lower limit for AES sensitivity since the Auger effect is a "three state" event necessitating at least three electrons. Neither H nor He can be detected with this technique. For K-level based transitions, Auger effects are dominant for $Z < 15$ while for L- and M-level transitions, AES data can be measured for $Z \geq 50$. The yield limits effectively prescribe a cutoff for AES sensitivity, but complex techniques can be utilised to identify heavier elements, such as uranium and americium, using the Auger effect.

Another critical quantity that determines yield of Auger electrons at a detector is the electron impact cross-section. Early approximations (in cm^2) of the cross-section were based on the work of Worthington and Tomlin,

$$\sigma_{ax}(E) = 1.3 \times 10^{13} b \frac{C}{E_p}$$

with b acting as a scaling factor between 0.25 and 0.35, and C a function of the primary electron beam energy, E_p. While this value of σ_{ax} is calculated for an isolated atom, a simple modification can be made to account for matrix effects:

$$\sigma(E) = \sigma_{ax}[1 + r_m(E_p,\alpha)]$$

where α is the angle to the surface normal of the incident electron beam; r_m can be established empirically and encompasses electron interactions with the matrix such as ionization due to backscattered electrons. Thus the total yield can be written as:

$$Y(t) = N_x \times \delta t \times \sigma(E,t)[1 - \omega_X]\exp(-t\cos\frac{\theta}{\lambda}) \times I(t) \times T \times \frac{d(\Omega)}{4\pi}$$

Here N_x is the number of x atoms per volume, λ the electron escape depth, θ the analyser angle, T the transmission of the analyser, $I(t)$ the electron excitation flux at depth t, dΩ the solid angle, and ät is the thickness of the layer being probed. Encompassed in these terms, especially the Auger yield which is related to the transition probability, is the quantum mechanical overlap of the initial and final state wave functions.

Precise expressions for the transition probability, based on first-order perturbation Hamiltonians, can be found in Thompson and Baker. Often, all of these terms are not known, so most analyses compare measured yields with external standards of known composition. Ratios of the acquired data to standards can eliminate common terms, especially experimental setup characteristics and material parameters, and can be used to determine element composition. Comparison techniques work best for samples of homogeneous binary materials or uniform surface layers, while elemental identification is best obtained from comparison of pure samples.

Uses and Limitations

There are a number of electron microscopes that have been specifically designed for use in Auger spectroscopy; these are termed scanning Auger microscopes (SAM) and can produce high resolution, spatially resolved chemical images. SAM images are obtained by stepping a focused electron beam across a sample surface and measuring the intensity of the Auger peak above the background of scattered electrons. The intensity map is correlated to a gray scale on a monitor with whiter areas corresponding to higher element concentration. In addition, sputtering is sometimes used with Auger spectroscopy to perform depth profiling experiments. Sputtering removes thin outer layers of a surface so that AES can be used to determine the underlying composition. Depth profiles are shown as either Auger peak height vs. sputter time or atomic concentration vs. depth. Precise depth milling through sputtering has made profiling an invaluable technique for chemical analysis of nanostructured materials and thin films. AES is also used extensively as an evaluation tool on and off fab lines in the microelectronics industry, while the versatility and sensitivity of the Auger process makes it a standard analytical tool in research labs.

Despite the advantages of high spatial resolution and precise chemical sensitivity attributed to AES, there are several factors that can limit the applicability of this technique, especially when evaluating solid specimens. One of the most common limitations encountered with Auger spectroscopy are charging effects in non-conducting samples. Charging results when the number of secondary electrons leaving the sample is different to the number of incident electrons, giving rise to a net polarity at the surface. Both positive and negative surface charges severely alter the yield of electrons emitted from the sample and hence distort the measured Auger peaks. To complicate matters, neutralisation methods employed in other surface analysis techniques, such as secondary ion mass spectrometry (SIMS), are not applicable to AES, as these methods usually involve surface bombardment with either electrons or ions (i.e. flood gun). Several processes have been developed to combat the issue of charging, though none of them is ideal and still make quantification of AES data difficult. One such technique involves depositing conductive pads near the analysis area to minimise regional charging. However, this type of approach limits SAM applications as well as the amount of sample material available for probing. A related technique involves thinning or "dimpling" a non-conductive layer with Ar^+ions and then mounting the sample to a conductive backing prior to AES. This method has been debated, with claims that the thinning process leaves elemental artifacts on a surface and/or creates damaged layers that distort bonding and promote chemical mixing in the sample.

As a result, the compositional AES data is considered suspect. The most common setup to minimise charging effects includes use of a glancing angle (~10°) electron beam and a carefully tuned bombarding energy (between 1.5 keV and 3 keV). Control of both the angle and energy can subtly alter the number of emitted electrons vis-à-vis the incident electrons and thereby reduce or altogether eliminate sample charging.

In addition to charging effects, AES data can be obscured by the presence of characteristic energy losses in a sample and higher order atomic ionization events. Electrons ejected from a solid will generally undergo multiple scattering events and lose energy in the form of collective electron density oscillations called plasmons. If plasmon losses have energies near that of an Auger peak, the less intense Auger process may become dwarfed by the plasmon peak.

As Auger spectra are normally weak and spread over many eV of energy, they are difficult to extract from the background and in the presence of plasmon losses, deconvolution of the two peaks becomes extremely difficult.

For such spectra, additional analysis through chemical sensitive surface techniques like x-ray photoelectron spectroscopy (XPS) is often required to disentangle the peaks. Sometimes an Auger spectrum can also exhibit "satellite" peaks at well-defined off-set energies from the parent peak. Origin of the satellites is usually attributed to multiple ionization events in an atom or ionization cascades in which a series of electrons is emitted as relaxation occurs for core holes of multiple levels. The presence of satellites can distort the true Auger peak and/or small peak shift information due to chemical bonding at the surface. Several studies have been undertaken to further quantify satellite peaks.

Despite these sometimes substantial drawbacks, Auger electron spectroscopy is a widely used surface analysis technique that has been successfully applied to many diverse fields ranging from gas phase chemistry to nanostructure characterisation. Very new class of high-resolving electrostatic energy analysers recently developed – the face-field analysers (FFA) can be used for remote electron spectroscopy of distant surfaces or surfaces with large roughness or even with deep dimples. These instruments are designed as if to be specifically used in combined scanning electron microscopes (SEMs). "FFA" in principle have no perceptible end-fields, which usually distort focusing in most of analysers known, for example, well known CMA.

Sensitivity, quantitative detail, and ease of use have brought AES from an obscure nuisance effect to a functional and practical characterisation technique in just over fifty years. With applications both in the research laboratory and industrial settings, AES will continue to be a cornerstone of surface-sensitive electron-based spectroscopies.

X-ray Photoelectron Spectroscopy

X-ray photoelectron spectroscopy (XPS) is a quantitative spectroscopic technique that measures the elemental composition, empirical formula, chemical state and electronic state of the elements that exist within a material. XPS spectra are obtained by irradiating a material with a beam of X-rays while simultaneously measuring the kinetic energy and number of electrons that escape from the top 1 to 10nm of the material being analysed. XPS requires ultra high vacuum (UHV) conditions.

XPS is a surface chemical analysis technique that can be used to analyse the surface chemistry of a material in its "as received" state, or after some treatment, for example: fracturing, cutting or scraping in air or UHV to expose the bulk chemistry, ion beam etching to clean off some of the surface contamination, exposure to heat to study the changes due to heating, exposure to reactive gases or solutions, exposure to ion beam implant, exposure to ultraviolet light.

XPS is also known as ESCA, an abbreviation for Electron Spectroscopy for Chemical Analysis. XPS detects all elements with an atomic number (Z) of 3 (lithium) and above. It cannot detecthydrogen (Z = 1) or helium (Z = 2) because the diametre of these orbitals is so small, reducing the catch probability to almost zero. Detection limits for most of the elements are in the parts per thousand range. Detection limits of parts per million (ppm) are possible, but require special conditions: concentration at top surface or very long collection time (overnight). XPS is routinely used to analyse inorganic compounds, metal alloys, semiconductors, polymers, elements, catalysts, glasses, ceramics, paints, papers, inks, woods, plant parts, make-up, teeth, bones, medical implants, bio-materials, viscous oils, glues, ion modified materials and many others.

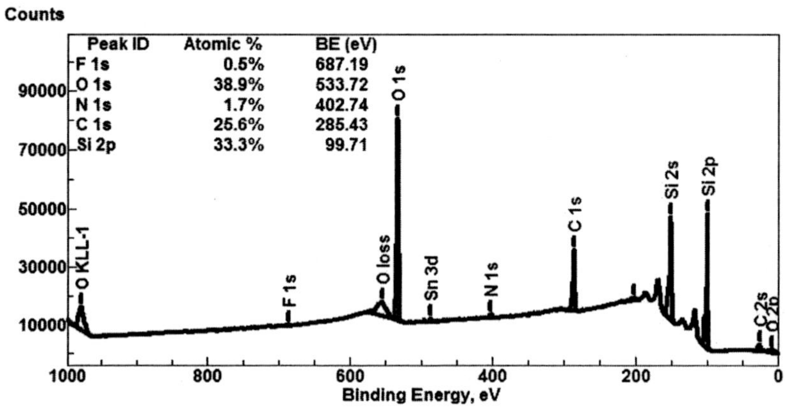

Figure: Wide-scan survey spectrum for all elements.

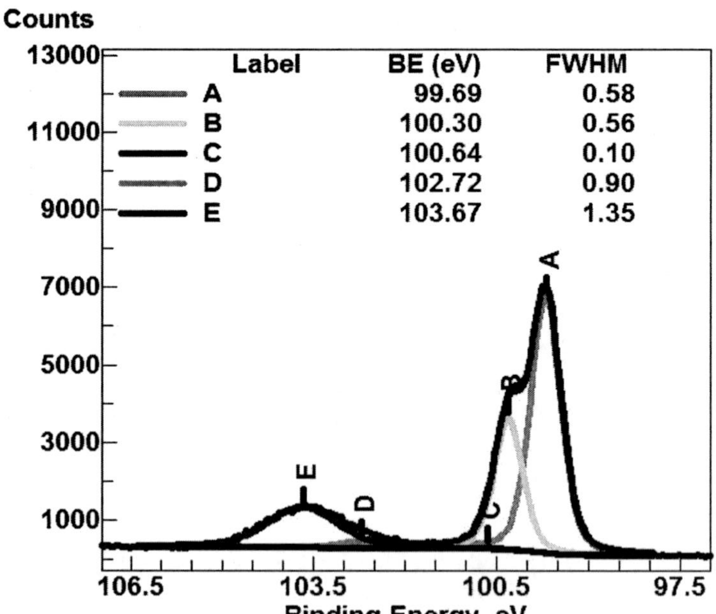

Figure: High-resolution spectrum for Si(2p) signal.

XPS is used to measure:

- Elemental composition of the surface (top 1–10 nm usually)
- Empirical formula of pure materials
- Elements that contaminate a surface
- Chemical or electronic state of each element in the surface
- Uniformity of elemental composition across the top surface (or line profiling or mapping)
- Uniformity of elemental composition as a function of ion beam etching (or depth profiling)

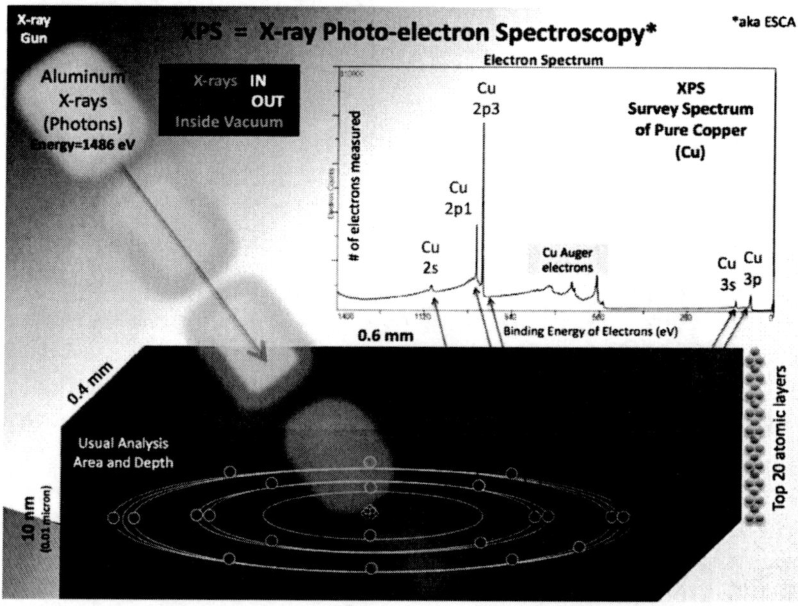

Figure: Rough schematic of XPS physics - "Photoelectric Effect.

XPS can be performed using either a commercially built XPS system, a privately built XPS system or a synchrotron-based light source combined with a custom designed electron analyser. Commercial XPS instruments in the year 2005 used either a highly focused 20 to 200 micrometre beam of monochromatic aluminium Kα X-rays or a broad 10–30 mm beam of non-monochromatic (polychromatic) magnesium X-rays. A few, special design, XPS instruments can analyse volatile liquids or gases, materials at low or high temperatures or materials at roughly 1 torr vacuum, but there are relatively few of these types of XPS systems. Because the energy of an X-ray with particular wavelength is known, the electron binding energy of each of the emitted electrons can be determined by using an equation that is based on the work of Ernest Rutherford (1914):

$$E_{\text{binding}} = E_{\text{photon}} - \left(E_{\text{kinetic}} + \phi\right)$$

where E_{binding} is the binding energy (BE) of the electron, E_{photon} is the energy of the X-ray photons being used, E_{kinetic} is the kinetic energy of the electron as measured by the instrument and φ is the work function of the spectrometre (not the material).

History

In 1887, Heinrich Rudolf Hertz discovered the photoelectric effect that was explained in 1905 by Albert Einstein (Nobel Prize in Physics 1921). Two years later, in 1907, P.D. Innes experimented with a Rφntgen tube, Helmholtz coils, a magnetic field hemisphere (electron energy analyser) and photographic plates to record broad bands of emitted electrons as a function of velocity, in effect recording the first XPS spectrum. Other researchers, Henry Moseley, Rawlinson and Robinson, independently performed various experiments trying to sort out the details in the broad bands. Wars halted research on XPS.

After WWII, Kai Siegbahn and his group in Uppsala (Sweden) developed several significant improvements in the equipment and in 1954 recorded the first high energy resolution XPS spectrum of cleaved sodium chloride (NaCl) revealing the potential of XPS. A few years later in 1967, Siegbahn published a comprehensive study on XPS bringing instant recognition of the utility of XPS. In cooperation with Siegbahn, Hewlett-Packard in the USA produced the first commercial monochromatic XPS instrument in 1969. Siegbahn received the Nobel Prize in 1981 to acknowledge his extensive efforts to develop XPS into a useful analytical tool.

In parallel with Siegbahn's work, David Turner at Imperial College (and later at Oxford) in the UK developed ultraviolet photoelectron spectroscopy (UPS) on molecular species using helium lamps.

Physics

A typical XPS spectrum is a plot of the number of electrons detected (sometimes per unit time) (Y-axis, ordinate) versus the binding energy of the electrons detected (X-axis, abscissa). Each element produces a characteristic set of XPS peaks at characteristic binding energy values that directly identify each element that exist in or on the surface of the material being analysed. These characteristic peaks correspond to the electron configuration of the electrons within the atoms, e.g., 1s, 2s, 2p, 3s, etc. The number of detected electrons in each of the characteristic peaks is directly related to the amount of element within the area (volume) irradiated. To generate atomic percentage values, each raw XPS signal must be corrected by dividing its signal intensity (number of electrons detected) by a "relative sensitivity factor" (RSF) and normalised over all of the elements detected.

To count the number of electrons at each kinetic energy value, with the minimum of error, XPS must be performed under ultra-high vacuum (UHV) conditions because electron counting detectors in XPS instruments are typically one metre away from the material irradiated with X-rays.

It is important to note that XPS detects only those electrons that have actually escaped into the vacuum of the instrument. The photo-emitted electrons that have escaped into the vacuum of the instrument are those that originated from within the top 10 to 12 nm of the material. All of the deeper photo-emitted electrons, which were generated as the X-rays penetrated 1– 5 micrometres of the material, are either recaptured or trapped in various excited states within the material. For most applications, it is, in effect, a non-destructive technique that measures the surface chemistry of any material.

Components of an XPS system

The main components of a commercially made XPS system include:

- A source of X-rays
- An ultra-high vacuum (UHV) stainless steel chamber with UHV pumps
- An electron collection lens
- An electron energy analyser
- Mu-metal magnetic field shielding
- An electron detector system
- A moderate vacuum sample introduction chamber
- Sample mounts
- A sample stage
- A set of stage manipulators.

Monochromatic aluminium K-alpha X-rays are normally produced by diffracting and focusing a beam of non-monochromatic X-rays off of a thin disc of natural, crystalline quartz with a <1010> orientation. The resulting wavelength is 8.3386 angstroms (0.83386 nm) which corresponds to a photon energy of 1486.7 eV. The energy width of the monochromated X-rays is 0.16 eV, but the common electron energy analyser (spectrometre) produces an ultimate energy resolution on the order of 0.25 eV which, in effect, is the ultimate energy resolution of most commercial systems. When working under practical, everyday conditions, high energy resolution settings will produce peak widths (FWHM) between 0.4–0.6 eV for various pure elements and some compounds. Non-monochromatic magnesium X-rays have a wavelength of 9.89 angstroms (0.989 nm) which corresponds to a photon energy of 1253 eV. The energy width of the non-monochromated X-ray is roughly 0.70 eV, which, in effect is the ultimate energy resolution of a system using non-monochromatic X-rays. Non-monochromatic X-ray sources do not use any crystals to diffract the X-rays which allows all primary X-rays lines and the full range of high energy Bremsstrahlung X-rays (1–12 keV) to reach the surface. The typical ultimate high energy resolution (FWHM) when using this source is 0.9–1.0 eV, which includes with the spectrometre-induced broadening, pass-energy settings and the peak-width of the non-monochromatic magnesium X-ray source.

Uses and Capabilities

XPS is routinely used to determine:

- What elements and the quantity of those elements that are present within the top 1-12 nm of the sample surface
- What contamination, if any, exists in the surface or the bulk of the sample
- Empirical formula of a material that is free of excessive surface contamination
- The chemical state identification of one or more of the elements in the sample
- The binding energy of one or more electronic states
- The thickness of one or more thin layers (1–8 nm) of different materials within the top 12 nm of the surface
- The density of electronic states.

Capabilities of Advanced Systems

- Measure uniformity of elemental composition across the top the surface (or *line profiling or mapping*)
- Measure uniformity of elemental composition as a function of depth by ion beam etching (or *depth profiling*)
- Measure uniformity of elemental composition as a function of depth by tilting the sample (or *angle resolved XPS*)

Chemical States from XPS Analyses

The ability to produce Chemical State information from the topmost 1-12 nm of any surface makes XPS a unique and invaluable tool for understanding the chemistry of any surface either, as received, or after physical or chemical treatment(s). Because modern systems use monochromatic X-ray sources, XPS measurements leave the surface free of any degradation with few exceptions.

Chemical state analysis of the surface of polymers readily reveals the presence or absence of the chemical states of carbon known as: carbide (C^{2-}), hydrocarbon (C-C), alcohol (C-OH), ketone (C=O), organic ester (COOR), carbonate (CO_3), fluoro-hydrocarbon (CF_2-CH_2), trifluorocarbon (CF_3).

Chemical state analysis of the surface of a silicon wafer readily reveals the presence or absence of the chemical states of silicon known as: n-doped silicon, p-doped silicon, silicon suboxide (Si_2O), silicon monoxide (SiO), Si_2O_3, silicon dioxide (SiO_2).

Routine Limits of XPS

Quantitative Accuracy

- XPS is widely used to generate empirical formula because it readily yields excellent quantitative accuracy from homogeneous solid state materials
- Quantitative accuracy depends on several parameters such as: signal-to-noise ratio, peak intensity, accuracy of relative sensitivity factors, correction for electron transmission function, surface volume homogeneity, correction for energy dependency of electron mean free path, and degree of sample degradation due to analysis.
- Under optimum conditions, the quantitative accuracy of the atomic percent (at%) values calculated from the Major XPS Peaks is 90-95% for each major peak. If a high level quality control protocol is used, the accuracy can be further improved.
- Under routine work conditions, where the surface is a mixture of contamination and expected material, the accuracy ranges from 80-90% of the value reported in atomic percent values.
- The quantitative accuracy for the weaker XPS signals, that have peak intensities 10-20% of the strongest signal, are 60-80% of the true value.

Analysis Time

- 1–10 minutes for a survey scan that measures the amount of all elements, 1– 10 minutes for high energy resolution scans that reveal chemical state differences, 1– 4 hours for a depth profile that measures 4– 5 elements as a function of etched depth (usual final depth is 1,000 nm)

Detection Limits

- 0.1–1.0 at% (0.1 at% = 1 part per thousand = 1000 ppm). (Ultimate detection limit for most elements is approximately 100 ppm, which requires 8–16 hours.)

Measured Area

- Measured area depends on instrument design. The minimum analysis area ranges from 10 to 200 micrometres. Largest size for a monochromatic beam of X-rays is 1–5 mm. Non-monochromatic beams are 10–50 mm in diametre. Spectroscopic image resolution levels of 200 nm or below has been achieved on latest imaging XPS instruments using synchrotron radiation as X-ray source.

Sample size Limits

Older instruments accept samples: 1×1 to 3×3 cm. Present systems can accept samples up to 30×30 cm.

Degradation During Analysis

Depends on the sensitivity of the material to the wavelength of X-rays used, the total dose of the X-rays, the temperature of the surface and the level of the vacuum. Metals, alloys, ceramics and most glasses are not measurably degraded by either non-monochromatic or monochromatic X-rays. Some, but not all, polymers, catalysts, certain highly oxygenated compounds, various inorganic compounds and fine organics are degraded by either monochromatic or non-monochromatic X-ray sources.

Non-monochromatic X-ray sources produce a significant amount of high energy Bremsstrahlung X-rays (1– 15 keV of energy) which directly degrade the surface chemistry of various materials. Non-monochromatic X-ray sources also produce a significant amount of heat (100 to 200 °C) on the surface of the sample because the anode that produces the X-rays is typically only 1 to 5 cm (2 in) away from the sample. This level of heat, when combined with the Bremsstrahlung X-rays, acts synergistically to increase the amount and rate of degradation for certain materials. Monochromatic X-ray sources, because they are far away (50– 100 cm) from the sample, do not produce any heat effects. Monochromatic X-ray sources are monochromatic because the quartz monochromator system diffracted the Bremsstrahlung X-rays out of the X-ray beam which means the sample only sees one X-ray energy, for example: 1.486 keV if aluminium K-alpha X-rays are used.

Because the vacuum removes various gases (e.g. O_2, CO) and liquids (e.g. water, alcohol, solvents) that were initially trapped within or on the surface of the sample, the chemistry and morphology of the surface will continue to change until the surface achieves a steady state. This type of degradation is sometimes difficult to detect.

Materials Routinely Analysed by XPS

Inorganic compounds, metal alloys, semiconductors, polymers, pure elements, catalysts, glasses, ceramics, paints, papers, inks, woods, plant parts, make-up, teeth, bones, human implants, biomaterials, viscous oils, glues, ion modified materials

Organic chemicals are not routinely analysed by XPS because they are readily degraded by either the energy of the X-rays or the heat from non-monochromatic X-ray sources.

Charge Compensation Techniques

- Low-voltage electron beam (1-20 eV) (or electron flood gun)
- UV lights
- Low-voltage argon ion beam with low-voltage electron beam (1-10 eV)
- Aperture masks
- Mesh screen with low-voltage electron beams

Sample Preparation

- Sample handling
- Sample cleaning
- Sample mounting.

Data Processing

Peak Identification

The identification of peaks in any survey spectrum is possible because Prof. Kai Siegbahn improved the energy resolution of his XPS instrument to the point that observed signals were both tall and narrow with respect to the energy range measured (0-1400 eV). In the course of improving his XPS instruments, he collected spectra from pure conductive elements. That collection of spectra yielded peaks with energies (reported as BEs) that are characteristic for each specific element. The number of peaks produced by a single element varies from 1 to more than 20. Tables of binding energies (BEs) that identify the shell and spin-orbit of each peak produced by a given element are included with modern XPS instruments, and can be found in various handbooks [citations] and websites [citations]. Because these experimentally determined BEs are characteristic of specific elements, they can be directly used to identify experimentally measured peaks of a material with unknown elemental composition.

Before beginning the process of peak identification, the analyst must determine if the BEs of the unprocessed survey spectrum (0-1400 eV) have or have not been shifted due to a positive or negative surface charge. This is most often done by looking for two peaks that due to the presence of carbon and oxygen. {tbc}

Charge Referencing Insulators

Charge referencing is needed when a sample suffers either a positive (+) or negative (-) charge induced shift of experimental BEs. Charge referencing is needed to obtain meaningful BEs from both wide-scan, high sensitivity (low energy resolution) survey spectra (0-1100 eV), and also narrow-scan, chemical state (high energy resolution) spectra.

Charge induced shifting causes experimentally measured BEs of XPS peaks to appear at BEs that are greater or smaller than true BEs. Charge referencing is performed by adding or subtracting a "Charge Correction Factor" to each of the experimentally measured BEs. In general, the BE of the hydrocarbon peak of the C (1s) XPS signal is used to charge reference (charge correct) all BEs obtained from non-conductive (insulating) samples or conductors that have been deliberately insulated from the sample mount.

Charge induced shifting is normally due to: a modest excess of low voltage (-1 to -20 eV) electrons attached to the surface, or a modest shortage of electrons (+1 to +15 eV) within the top 1-12 nm of the sample caused by the loss of photo-emitted electrons. The degree of charging depends on various factors. If, by chance, the charging of the surface is excessively positive, then the spectrum might appear as a series of rolling hills, not sharp peaks as shown in the example spectrum.

The C (1s) BE of the hydrocarbon species (moieties) of the "Adventitious" carbon that appears on all, air-exposed, conductive and semi-conductive materials is normally found between 284.5 eV and 285.5 eV. For convenience, the C (1s) of hydrocarbon moieties is defined to appear between 284.6 eV and 285.0 eV. A value of 284.8 eV has become popular in recent years. However, some recent reports indicate that 284.9 eV or 285.0 eV represents hydrocarbons attached on metals, not the natural native oxide. The 284.8 eV BE is routinely used as the "Reference BE" for charge referencing insulators. When the C (1s) BE is used for charge referencing, then the charge correction factor is the difference between 284.8 eV and the experimentally measured C (1s) BE of the hydrocarbon moieties.

When using a monochromatic XPS system together with a low voltage electron flood gun for charge compensation the experimental BEs of the C (1s) hydrocarbon peak is often 4-5 eV smaller than the reference BE value (284.8 eV). In this case, all experimental BEs appear at lower BEs than expected and need to be increased by adding a value ranging from 4 to 5 eV. Non-monochromatic XPS systems are not usually equipped with a low voltage electron flood gun so the BEs will normally appear at higher BEs than expected. It is normal to subtract a charge correction factor from all BEs produced by a non-monochromatic XPS system.

Conductive materials and most native oxides of conductors should never need charge referencing. Conductive materials should never be charge referenced unless the topmost layer of the sample has a thick non-conductive film.

Peak-fitting

The process of peak-fitting high energy resolution XPS spectra is still a mixture of art, science, knowledge and experience. The peak-fit process is affected by instrument design, instrument components, experimental settings (aka analysis conditions) and sample variables. Most instrument parameters are constant while others depend on the choice of experimental settings.

Before starting any peak-fit effort, the analyst performing the peak-fit needs to know if the topmost 15 nm of the sample is expected to be a homogeneous material or is expected to be a mixture of materials. If the top 15 nm is a homogeneous material with only very minor amounts of adventitious carbon and adsorbed gases, then the analyst can use theoretical peak area ratios to enhance the peak-fitting process.

Variables that affect or define peak-fit results include:

- FWHMs
- Chemical Shifts
- Peakshapes
- Instrument design factors
- Experimental settings
- Sample factors.

FWHMs

When using high energy resolution experiment settings on an XPS equipped with a monochromatic Al K-alpha X-ray source, the FWHM of the major XPS peaks range from 0.3 eV to 1.7 eV. The following is a simple summary of FWHM from major XPS signals:

Main metal peaks (e.g. 1s, 2p3, 3d5, 4f7) from pure metals have FWHMs that range from 0.30 eV to 1.0 eV

Main metal peaks (e.g. 1s, 2p3, 3d5, 4f7) from binary metal oxides have FWHMs that range from 0.9 eV to 1.7 eV

The O (1s) peak from binary metal oxides have FWHMs that, in general, range from 1.0 eV to 1.4 eV

The C (1s) peak from adventitious hydrocarbons have FWHMs that, in general, range from 1.0 eV to 1.4 eV

Chemical Shifts

Chemical shift values depend on the degree of electron bond polarisation between nearest neighbour atoms. A specific chemical shift is the difference in BE values of one specific chemical stateversus the BE of the pure element.

Peaks derived from peak-fitting a raw chemical state spectrum are due to the presence of different chemical states.

Peakshapes

Depends on instrument parameters, experimental parameters and sample characteristics

Instrument Design Factors

FWHM and purity of X-rays used (monochromatic Al, non-monochromatic Mg, Synchrotron, Ag, Zr...)

Design of electron analyser (CMA, HSA, retarding field...)

Experiment Settings

Settings of the electron analyser (e.g. pass energy, step size)

Sample Factors

Physical form of the sample (single crystal, polished, powder, corroded...)

Number of physical defects within the analysis volume (from Argon ion etching, from laser cleaning...)

Photoemission Spectroscopy

Photoemission spectroscopy (PES), also known as photoelectron spectroscopy, refers to energy measurement of electrons emitted from solids, gases or liquids by the photoelectric effect, in order to determine the binding energies of electrons in a substance. The term refers to various techniques, depending on whether the ionization energy is provided by an X-ray photon, an EUV photon, or an ultraviolet photon. Regardless of the incident photon beam however, all photoelectron spectroscopy revolves around the general theme of surface analysis by measuring the ejected electrons.

X-ray photoelectron spectroscopy (XPS) was developed by Kai Siegbahn starting in 1957 and is used to study the energy levels of atomic core electrons, primarily in solids. Siegbahn referred to the technique as Electron Spectroscopy for Chemical Analysis (ESCA), since the core levels have small chemical shifts depending on the chemical environment of the atom which is ionized, allowing chemical structure to be determined. Siegbahn was awarded the Nobel Prize in 1981 for this work. XPS is sometimes referred to as PESIS (photoelectron spectroscopy for inner shells) whereas the lower energy radiation of uv-light is referred to as PESOS (outer shells) because it cannot excite core electrons.

In the ultraviolet region, the method is usually referred to as photoelectron spectroscopy for the study of gases, and photoemission spectroscopy for solid surfaces.

Ultraviolet Photoelectron Spectroscopy

Ultra-violet photoelectron spectroscopy (UPS) is used to study valence energy levels and chemical bonding; especially the bonding character of molecular orbitals. The method was developed originally for gas-phase molecules in 1962 by David W. Turner, and other early workers included David C.Frost, J.H.D. Eland and K. Kimura. Later, Richard Smalley modified the technique and used a UV laser to excite the sample, in order to measure the binding energy of electrons in gaseous molecular clusters.

Extreme ultraviolet photoelectron spectroscopy (EUPS) lies in between XPS and UPS. It is typically used to assess the valence band structure. Compared to XPS it gives better energy resolution, and compared to UPS the ejected electrons are faster, resulting in a better spectrum signal.

Physical Principle

The physics behind the PES technique is an application of the photoelectric effect. The sample is exposed to a beam of UV or XUV light inducing photoelectric ionization.

The energies of the emitted photoelectrons are characteristic of their original electronic states, and depend also on vibrational state and rotational level. For solids, photoelectrons can escape only from a depth on the order of nanometres, so that it is the surface layer which is analysed.

Because of the high frequency of the light, and the substantial charge and energy of emitted electrons, photoemission is one of the most sensitive and accurate techniques for measuring the energies and shapes of electronic states and molecular and atomic orbitals. Photoemission is also among the most sensitive methods of detecting substances in trace concentrations, provided the sample is compatible with ultra-high vacuum and the analyte can be distinguished from background.

Typical PES (UPS) instruments use helium gas sources of UV light, with photon energy up to 52 eV (corresponding to wavelength 23.7 nm). The photoelectrons that actually escaped into the vacuum are collected, energy resolved, slightly retarded and counted, which results in a spectrum of electron intensity as a function of the measured kinetic energy. Because binding energy values are more readily applied and understood, the kinetic energy values, which are source dependent, are converted into binding energy values, which are source independent. This is achieved by applying Einstein's relation E_k = hí “ E_B. The hí term of this equation is due to the energy (frequency) of the UV light that bombards the sample. Photoemission spectra are also measured using synchrotron radiation sources.

The binding energies of the measured electrons are characteristic of the chemical structure and molecular bonding of the material. By adding a source monochromator and increasing the energy resolution of the electron analyser, peaks appear with full width at half maximum (FWHM) less than 5–8 meV.

Ultraviolet photoelectron spectroscopy (UPS) refers to the measurement of kinetic energy spectra of photoelectrons emitted by molecules which have absorbed ultraviolet photons, in order to determine molecular energy levels in the valence region.

Basic Theory

If Einstein's photoelectric law is applied to a free molecule, the kinetic energy (E_K) of an emitted photoelectron is given by

$$E_K = h\nu - I ,$$

where h is Planck's constant, í is the frequency of the ionizing light, and I is an ionization energy corresponding to the energy of an occupied molecular orbital.

History

Prior to 1960, virtually all measurements of photoelectron kinetic energies were for electrons emitted from metals and other solid surfaces. About 1956 Kai Siegbahn developed X-ray photoelectron spectroscopy (XPS) for surface chemical analysis. This method uses x-ray sources to study energy levels of atomic core electrons, and at the time had an energy resolution of about 1 eV (electronvolt).

The ultraviolet method (UPS) was developed to study the photoelectron spectra of free molecules in the gas phase by David W. Turner, a physical chemist at Imperial College in London and then at Oxford University, in a series of publications from 1962 to 1967. As a photon source, he used a helium discharge lamp which emits a wavelength of 58.4 nm (corresponding to an energy of 21.2 eV) in the vacuum ultraviolet region. With this source Turner's group obtained an energy resolution of 0.02 eV. Turner referred to the method as "molecular photoelectron spectroscopy", now usually "Ultraviolet photoelectron spectroscopy" or UPS. As compared to XPS, UPS is limited to energy levels of valence electrons, but measures them more accurately. After 1967 commercial UPS spectrometres became available.

Application

The UPS measures experimental molecular orbital energies for comparison with theoretical values from quantum chemistry, which was also extensively developed in the 1960s. The photoelectron spectrum of a molecule contains a series of peaks each corresponding to one valence-region molecular orbital energy level. Also, the high resolution allowed the observation of fine structure due to vibrational levels of the molecular ion, which facilitates the assignment of peaks to bonding, nonbonding or antibonding molecular orbitals. The method was later extended to the study of solid surfaces where it is usually described as photoemission spectroscopy (PES). It is particularly sensitive to the surface region (to 10 nm depth), due to the short range of the emitted photoelectrons (compared to X-rays). It is therefore used to study adsorbed species and their binding to the surface, as well as their orientation on the surface.

A useful result from characterisation of solids by UPS is the determination of the work function of the material. An example of this determination is given by Park et al. Briefly, the full width of the photoelectron spectrum (from the highest kinetic energy/lowest binding energy point to the low kinetic energy cutoff) is measured and subtracted from the photon energy of the exciting radiation, and the difference is the work function. Often, the sample is electrically biased negative to separate the low energy cutoff from the spectrometre response.

Thermal Infrared Spectroscopy

Thermal infrared spectroscopy (TIR spectroscopy) is the subset of infrared spectroscopy that deals with radiation emitted in the infrared part of the electromagnetic spectrum. The emitted infrared radiation, though similar to blackbody radiation, is different in that the radiation is banded at characteristic vibrations in the material. The method measures the thermal infrared radiation emitted (as opposed to being transmitted or reflected) from a volume or surface. This method is commonly used to identify the composition of surface by analysing its spectrum and comparing it to previously measured materials. It is particularly suited to airborne and spaceborne applications.

Thermal Infrared Spectrometres

Airborne

- TIMS: a multispectral radiometre flown on C-130, ER-2, and the Stennis Learjet aircraft.
- SEBASS: a hyperspectral sensor developed and operated by The Aerospace Corporation.
- OWL: a hyperspectral thermal infrared camera developed by Specim.

5

Optical Frameworks

Coma (Optics)

In optics (especially telescopes), the coma (aka *comatic aberration*) in an optical system refers to aberration inherent to certain optical designs or due to imperfection in the lens or other components which results in off-axis point sources such as stars appearing distorted, appearing to have a tail (coma) like a comet.

Specifically, coma is defined as a variation in magnification over the entrance pupil. In refractive or diffractive optical systems, especially those imaging a wide spectral range, coma can be a function of wavelength, in which case it is a form of chromatic aberration.

Coma is an inherent property of telescopes using parabolic mirrors. Light from a point source (such as a star) in the centre of the field is perfectly focused at the focal point of the mirror (unlike a spherical mirror, where light from the outer part of the mirror focuses closer to the mirror than light from the centre—spherical aberration). However, when the light source is off-centre (off-axis), the different parts of the mirror do not reflect the light to the same point. This results in a point of light that is not in the centre of the field looking wedge-shaped. The further off-axis, the worse this effect is. This causes stars to appear to have a cometary coma, hence the name.

Schemes to reduce spherical aberration without introducing coma include Schmidt, Maksutov, ACF and Ritchey-Chrétien optical systems. Correction lenses for Newtonian reflectors have been designed which reduce coma in telescopes below f/6. These work by means of a dual lens system of a plano-convex and a plano-concave lens fitted into an eyepiece adaptor which superficially resembles a Barlow lens.

Coma of a single lens or a system of lenses can be minimised (and in some cases eliminated) by choosing the curvature of the lens surfaces to match the application. Lenses in which both spherical aberration and coma are minimised at a single wavelength are called *bestform* or*aplanatic* or aspheric lenses. Vertical coma is the most common higher-order aberration in the eyes of patients with keratoconus

Astigmatism

An optical system with astigmatism is one where rays that propagate in two perpendicular planes have different foci. If an optical system with astigmatism is used to form an image of a cross, the vertical and horizontal lines will be in sharp focus at two different distances. The term comes from the Greek á- (*a*-) meaning "without" and σôßγμá (*stigma*), "a mark, spot, puncture".

Forms of Astigmatism

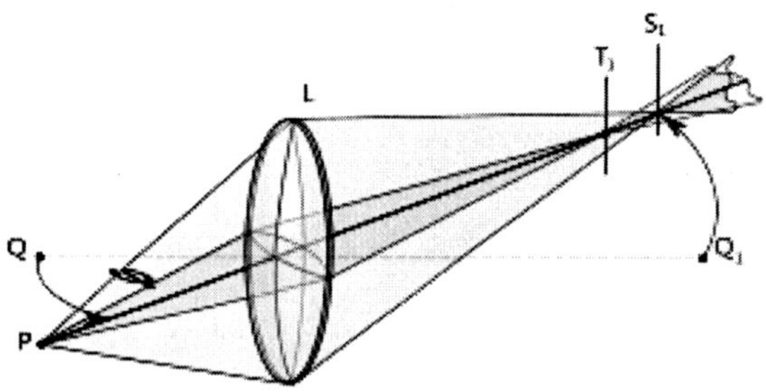

Figure: Astigmatism

There are two distinct forms of astigmatism. The first is a third-order aberration, which occurs for objects (or parts of objects) away from the optical axis. This form of aberration occurs even when the optical system is perfectly symmetrical. This is often referred to as a "monochromatic aberration", because it occurs even for light of a single wavelength. This terminology may be misleading, however, as the *amount* of aberration can vary strongly with wavelength in an optical system.

The second form of astigmatism occurs when the optical system is not symmetric about the optical axis. This may be by design (as in the case of a cylindrical lens), or due to manufacturing error in the surfaces of the components or misalignment of the components. In this case, astigmatism is observed even for rays from on-axis object points. This form of astigmatism is extremely important in vision science and eye care, since the human eye often exhibits this aberration due to imperfections in the shape of the cornea or the lens.

Third-order Astigmatism

In the analysis of this form of astigmatism, it is most common to consider rays from a given point on the object, which propagate in two special planes. The first plane is the *tangential plane*. This is the plane which includes both the object point being considered and the axis of symmetry. Rays that propagate in this plane are called tangential rays. Planes that include the optical axis are *meridional* planes. It is common to simplify problems in radially-symmetric optical systems by choosing object points in the vertical ("y") plane only. This plane is then sometimes referred to as *the* meridional plane.

The second special plane is the *sagittal plane*. This is defined as the plane, orthogonal to the tangential plane, which contains the object point being considered and intersects the optical axis at the entrance pupil of the optical system. This plane contains the chief ray, but does not contain the optic axis. It is therefore a *skew* plane, in other words not a meridional plane. Rays propagating in this plane are called sagittal rays.

In third-order astigmatism, the sagittal and transverse rays form foci at different distances along the optic axis. These foci are called the *sagittal focus* and the *transverse focus*, respectively. In the presence of astigmatism, an off-axis point on the object is not sharply imaged by the optical system. Instead, sharp *lines* are formed at the sagittal and transverse foci.

The image at the transverse focus is a short line, oriented in the direction of the *sagittal* plane; images of circles centred on the optic axis, or lines tangential to such circles, will be sharp in this plane. The image at the sagittal focus is a short line, oriented in the *tangential* direction; images of spokes radiating from the centre are sharp at this focus. In between these two foci, a round but "blurry" image is formed. This is called the *medial focus* or *circle of least confusion*. This plane often represents the best compromise image location in a system with astigmatism. The amount of aberration due to astigmatism is proportional to the square of the angle between the rays from the object and the optical axis of the system. With care, an optical system can be designed to reduce or eliminate astigmatism. Such systems are called anastigmats.

Astigmatism in Systems that are not Rotationally Symmetric

If an optical system is not axisymmetric, either due to an error in the shape of the optical surfaces or due to misalignment of the components, astigmatism can occur even for on-axis object points. This effect is often used deliberately in complex optical systems, especially certain types of telescope.

In the analysis of these systems, it is common to consider tangential rays (as defined above), and rays in a meridional plane (a plane containing the optic axis) perpendicular to the tangential plane. This plane is called either the *sagittal meridional plane* or, confusingly, just the *sagittal plane*.

Ophthalmic Astigmatism

In optometry and ophthalmology, the vertical and horizontal planes are identified as *tangential* and sagittal meridians, respectively. Ophthalmic astigmatism is a refraction error of the eye in which there is a difference in degree of refraction in different meridians. It is typically characterised by an aspherical, non-figure of revolution cornea in which the corneal profile slope and refractive power in one meridian is less than that of the perpendicular axis.

Astigmatism causes difficulties in seeing fine detail. Astigmatism can be often corrected by glasses with a lens that has different radii of curvature in different planes (a *cylindrical* lens), contact lenses, or refractive surgery.

Astigmatism is quite common. Studies have shown that about one in three people suffers from it. The prevalence of astigmatism increases with age. Although a person may not notice mild astigmatism, higher amounts of astigmatism may cause blurry vision, squinting, asthenopia, fatigue, or headaches.

There are a number of tests used by ophthalmologists and optometrists during eye examinations to determine the presence of astigmatism and to quantify the amount and axis of the astigmatism. ASnellen chart or other eye chart may initially reveal reduced visual acuity. A keratometre may be used to measure the curvature of the steepest and flattest meridians in the cornea's front surface. Corneal topography may also be used to obtain a more accurate representation of the cornea's shape. An autorefractor or retinoscopy may provide an objective estimate of the eye's refractive error and the use of Jackson cross cylinders in a phoropter may be used to subjectively refine those measurements. An alternative technique with the phoropter requires the use of a "clock dial" or "sunburst" chart to determine the astigmatic axis and power.

Astigmatism due to Misaligned or Malformed Lenses and Mirrors

Grinding and polishing of precision optical parts, either by hand or machine, typically employs significant downward pressure, which in turn creates significant frictional side pressures during polishing strokes that can combine to locally flex and distort the parts.

These distortions generally do not possess figure-of-revolution symmetry and are thus astigmatic, and slowly become permanently polished into the surface if the problems causing the distortion are not corrected. Astigmatic, distorted surfaces potentially introduce serious degradations in optical system performance.

Surface distortion due to grinding or polishing increases with the aspect ratio of the part (diametre to thickness ratio). To a first order, glass strength increases as the cube of the thickness. Thick lenses at 4:1 to 6:1 aspect ratios will flex much less than high aspect ratio parts, such as optical windows, which can have aspect ratios of 15:1 or higher.

The combination of surface or wavefront error precision requirements and part aspect ratio drives the degree of back support uniformity required, especially during the higher down pressures and side forces during polishing. Optical working typically involves a degree of randomness that helps greatly in preserving figure-of-revolution surfaces, provided the part is not flexing during the grind/polish process.

Deliberate Astigmatism in Optical Systems

Compact disc players use an astigmatic lens for focusing. When one axis is more in focus than the other, dot-like features on the disc project to oval shapes. The orientation of the oval indicates which axis is more in focus, and thus which direction the lens needs to move. A square arrangement of only four sensors can observe this bias and use it to bring the read lens to best focus, without being fooled by oblong pits or other features on the disc surface.

Some telescopes use deliberately astigmatic optics.

Petzval Field Curvature

Petzval field curvature, named for Joseph Petzval, describes the optical aberration in which a flat object normal to the optical axis (or a non-flat object past the hyperfocal distance) cannot be brought into focus on a flat image plane.

Consider an "ideal" single-element lens system for which all planar wave fronts are focused to a point a distance f from the lens. Placing this lens the distance f from a flat image sensor, image points near the optical axis will be in perfect focus, but rays off axis will come into focus before the image sensor, dropping off by the cosine of the angle they make with the optical axis. This is less of a problem when the imaging surface is spherical, as in the human eye.

Most current photographic lenses are designed to minimise field curvature, and so effectively have a focal length that increases with ray angle. However, film cameras could bend their image planes to compensate, particularly when the lens was fixed and known. This also included plate film, which could still be bent slightly. Digital sensors generally cannot be bent, though large mosaics of sensors (necessary anyway due to limited chip sizes) can be shaped to simulate a bend over larger scales.

Figure: The image-sensor array of the Kepler space observatory is curved to compensate for telescope's Petzval curvature.

The Petzval field curvature is equal to the Petzval sum over an optical system,

$$\sum_i \frac{n_{i+1} - n_i}{r_i n_{i+1} n_i}$$

where r_i is the radius of the ith surface and the ns are the indices of refraction on the first and second side of the surface.

Calibrated

Calibrated systems work from a table of lens/camera transfer functions:

- PTlens is a Photoshop plugin or standalone application which corrects complex distortion. It not only corrects for linear distortion, but also second degree and higher nonlinear components.
- DxO Labs' Optics Pro can correct complex distortion, and takes into account the focus distance.
- The Micro Four Thirds system cameras and lenses perform automatic distortion correction using correction parameters that are stored in each lens's firmware, and are applied automatically by the camera and RAW converter software. The optics of most of these lenses feature substantially more distortion than their counterparts in systems that don't offer such automatic corrections, but the software-corrected final images show noticeably less distortion than competing designs.

Manual

Manual systems allow manual adjustment of distortion parameters:

- Photoshop CS2 and Photoshop Elements (from version 5) include a manual Lens Correction filter for simple (pincushion/barrel) distortion
- Corel Paint Shop Pro Photo include a manual Lens Distortion effect for simple (barrel, fisheye, fisheye spherical and pincushion) distortion.
- The GIMP includes manual lens distortion correction (from version 2.4).
- PhotoPerfect has interactive functions for general pincushion adjustment, and for fringe (adjusting the size of the red, green and blue image parts).
- Hugin can be used to correct distortion, though that is not its primary application.

Related Phenomena

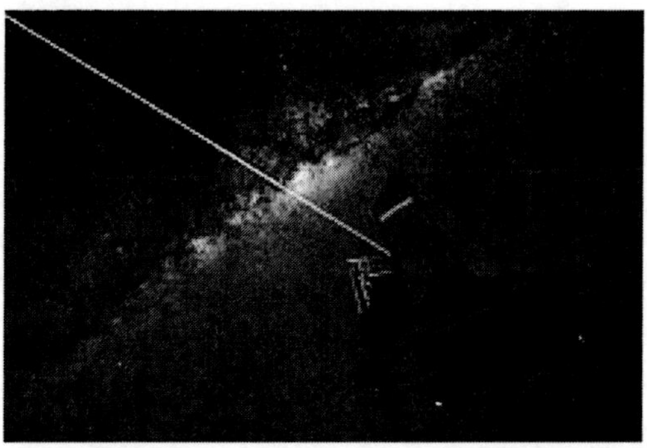

Figure: In astronomy, distortion caused by atmospheric turbulence is corrected by using adaptive optics.

Radial distortion is a failure of a lens to be rectilinear: a failure to image lines into lines. If a photograph is not taken straight-on then, even with a perfect rectilinear lens, rectangles will appear as trapezoids: lines are imaged as lines, but the angles between them are not preserved (tilt is not a conformal map). This effect can be controlled by using a perspective control lens, or corrected in post-processing.

Due to perspective, cameras image a cube as a squarefrustum (a truncated pyramid, with trapezoidal sides)—the far end is smaller than the near end. This creates perspective, and the rate at which this scaling happens (how quickly more distant objects shrink) creates a sense of a scene being deep or shallow. This cannot be changed or corrected by a simple transform of the resulting image, because it requires 3D information, namely the depth of objects in the scene. This effect is known as perspective distortion; the image itself is not distorted, but is perceived as distorted when viewed from a normal viewing distance. Note that if the centre of the image is closer than the edges (for example, a straight-on shot of a face), then barrel distortion and wide-angle distortion (taking the shot from close) both increase the size of the centre, while pincushion distortion and telephoto distortion (taking the shot from far) both decrease the size of the centre. However, radial distortion bends straight lines (out or in), while perspective distortion does not bend lines, and these are distinct phenomena. Fisheye lenses are wide-angle lenses with heavy barrel distortion and thus exhibit *both* these phenomena, so objects in the centre of the image (if shot from a short distance) are particularly enlarged: even if the barrel distortion is corrected, the resulting image is still from a wide-angle lens, and will still have a wide-angle perspective.

Chromatic Aberration

In optics, chromatic aberration (CA, also called achromatism or chromatic distortion) is a type of distortion in which there is a failure of a lens to focusall colours to the same convergence point. It occurs because have a different refractive index for different wavelengths of light (the dispersion of the lens). The refractive index decreases with increasing wavelength. Chromatic aberration manifests itself as "fringes" of colour along boundaries that separate dark and bright parts of the image, because each colour in the optical spectrum cannot be focused at a single common point. Since the focal length f of a lens is dependent on the refractive index n, different wavelengths of light will be focused on different positions.

Types

There are two types of chromatic aberration, axial (longitudinal), and transverse (lateral). Axial aberration occurs when different wavelengths of light are focused at different distances from the lens, i.e. different points on the optical axis (focus shift). Transverse aberration occurs when different wavelengths are focused at different positions in the focal plane (because the magnification and/or distortion of the lens also varies with wavelength; indicated in graphs as (change in) focus length). The acronym LCA is used, but ambiguous, and may refer to either longitudinal or lateral CA; for clarity, this article uses "axial" (shift in the direction of the optical axis) and "transverse" (shift perpendicular to the optical axis, in the plane of the sensor or film).

These two types have different characteristics, and may occur together. Axial CA occurs throughout the image, and is reduced by stopping down (this increases depth of field, so though the different wavelength focus at different distances, they are still in acceptable focus). Transverse CA does not occur in the centre, and increases towards the edge, but is not affected by stopping down.

In digital sensors, axial CA results in the red and blue planes being defocused (assuming that the green plane is in focus), which is relatively difficult to remedy in post-processing, while transverse CA results in the red, green, and blue planes being at different magnifications (magnification changing along radii, as in geometric distortion), and can be corrected by scaling the planes appropriately so they line up.

Minimisation

In the earliest uses of lenses, chromatic aberration was reduced by increasing the focal length of the lens where possible. For example, this could result in extremely long telescopes such as the very long aerial telescopes of the 17th century. Isaac Newton's theories about white light being composed of a spectrum of colours led him to the conclusion that uneven refraction of light caused chromatic aberration (leading him to build the first reflecting telescope, his Newtonian telescope, in 1668).

There exists a point called the circle of least confusion, where chromatic aberration can be minimised. It can be further minimised by using an achromatic lens or achromat, in which materials with differing dispersion are assembled together to form a compound lens. The most common type is an achromatic doublet, with elements made of crown and flint glass. This reduces the amount of chromatic aberration over a certain range of wavelengths, though it does not produce perfect correction.

By combining more than two lenses of different composition, the degree of correction can be further increased, as seen in an apochromatic lens or apochromat. Note that "achromat" and "apochromat" refer to the type of correction (2 or 3 wavelengths correctly focused), not the degree (how defocused the other wavelengths are), and an achromat made with sufficiently low dispersion glass can yield significantly better correction than an achromat made with more conventional glass. Similarly, the benefit of apochromats is not simply that they focus 3 wavelengths sharply, but that their error on other wavelength is also quite small.

Many types of glass have been developed to reduce chromatic aberration. These are low dispersion glass, most notably, glasses containing fluorite. These hybridized glasses have a very low level of optical dispersion; only two compiled lenses made of these substances can yield a high level of correction.

The use of achromats was an important step in the development of the optical microscope and intelescopes. An alternative to achromatic doublets is the use of diffractive optical elements. Diffractive optical elements have complementary dispersion characteristics to that of optical glasses and plastics. In the visible part of the spectrum, diffractives have an Abbe number of "3.5. Diffractive optical elements can be fabricated using diamond turning techniques.

Mathematics of Chromatic Aberration Minimisation

For a doublet consisting of two thin lenses in contact, the Abbe number of the lens materials is used to calculate the correct focal length of the lenses to ensure correction of chromatic aberration. If the focal lengths of the two lenses for light at the yellow Fraunhofer D-line (589.2 nm) are f_1 and f_2, then best correction occurs for the condition:

$$f_1 \cdot V_1 + f_2 \cdot V_2 = 0$$

where V_1 and V_2 are the Abbe numbers of the materials of the first and second lenses, respectively. Since Abbe numbers are positive, one of the focal lengths must be negative, i.e., a diverging lens, for the condition to be met.

The overall focal length of the doublet f is given by the standard formula for thin lenses in contact:

$$\frac{1}{f} = \frac{1}{f_1} + \frac{1}{f_2}$$

and the above condition ensures this will be the focal length of the doublet for light at the blue and red Fraunhofer F and C lines (486.1 nm and 656.3 nm respectively). The focal length for light at other visible wavelengths will be similar but not exactly equal to this. Chromatic aberration is used during an eye test to ensure that a correct lens power has been selected. The patient is confronted with red and green images and asked which is sharper. If the prescription is right, then the cornea, lens and prescribed lens will focus the red and green wavelengths just in front, and behind the retina, appearing of equal sharpness. If the lens is too powerful or weak, then one will focus on the retina, and the other will be much more blurred in comparison.

Image Processing to Reduce the Appearance of Lateral Chromatic Aberration

In some circumstances it is possible to correct some of the effects of chromatic aberration in digital post-processing. However in real-world circumstances, chromatic aberration results in permanent loss of some image detail. Detailed knowledge of the optical system used to produce the image can allow for some useful correction. In an ideal situation, post-processing to remove or correct lateral chromatic aberration would involve scaling the fringed colour channels, or subtracting some of a scaled versions of the fringed channels, so that all channels spatially overlap each other correctly in the final image.

As chromatic aberration is complex (due to its relationship to focal length, etc.) some camera manufacturers employ lens-specific chromatic aberration appearance minimisation techniques. All Nikon DSLRs with CMOS sensors and all Panasonic Lumix DSLRs, additionally some Nikon and Panasonic compact cameras, do such processing automatically in camera for JPEGimages.

Nikon DSLRs additionally store correction-data in RAW-files for use by Nikon Capture, View NX and some other RAW tools. Due to the Canon EF lens mount all-electronic lens communication and coupling system, the Canon's Digital Photo Professional software has one of the most advanced chromatic aberration appearance minimisation (also vignetting and distortion handling) systems that uses lens model-specific focusing distance, focal length and aperture information stored in the Canon RAW (CR2) files. Third party software tools such as PTLens are also capable of performing complex chromatic aberration appearance minimisation with their large database of cameras and lens.

In reality, even a theoretically perfect post-processing based chromatic aberration reduction-removal-correction systems do not increase image detail as a lens that is optically well corrected for chromatic aberration would for the following reasons:

- Rescaling is only applicable to lateral chromatic aberration but there is also longitudinal chromatic aberration
- Rescaling individual colour channels result in a loss of resolution from the original image
- Most camera sensors only capture a few and discrete (e.g., RGB) colour channels but chromatic aberration is not discreet and occurs across the light spectrum

- The dyes used in the digital camera sensors for capturing colour are not very efficient so cross-channel colour contamination is unavoidable and causes, for example, the chromatic aberration in the red channel to also be blended in to the green channel along with any green chromatic aberration.

A keratometre may be used to measure the curvature of the steepest and flattest meridians in the cornea's front surface. Corneal topography may also be used to obtain a more accurate representation of the cornea's shape.

An autorefractor or retinoscopy may provide an objective estimate of the eye's refractive error and the use of Jackson cross cylinders in a phoropter may be used to subjectively refine those measurements.

The above are closely related to the specific scene that is captured so no amount of programming and knowledge of the capturing equipment (e.g., camera and lens data) can overcome these limitations.

Bibliography

Abramowitz, Mortimer: *Contrast Methods in Microscopy: Transmitted Light*, Olympus America Inc., Melville, New York, 2017.

Adams, M.J.: *An Introduction to Optical Waveguides*, John Wiley and Sons, New York, 2012.

Agrawal, G.P.: *Nonlinear Fiber Optics*, Academic Press, Inc., New York, 2015.

Agrawal, I.D., and G. Lu: *Flouride Glass Fiber Optics*, Academic Press, Inc., New York, 2011.

Alan E. Willner, *Optical Fiber Communication Systems*, Norwood, MA, USA: Artech House, 2016.

Allard, F.C.: *Fiber Optics Handbook*, McGraw-Hill, Inc., New York, 2020.

Amann, M-C. and J. Buus: *Tunable Laser Diodes,* Artech House, Boston, 2018.

Arnaud, J.A.: *Beam and Fiber Optics*, Academic Press, Inc., New York, 2016.

Barnard, Joseph Edwin: *Practical Photo-Micrography.*, Edward Arnold and Company, London, United Kingdom, 2013.

Bass, M. : *Handbook of Optics*, McGraw-Hill, New York, 2015.

Bausch, Edward: *Manipulation of the Microscope.*, Bausch and Lomb Optical Company, Rochester, New York, 2011.

Bausch, Edward: *Manipulation of the Microscope.*, Bausch and Lomb Optical Company, Rochester, New York, 2011.

Baxes, Gregory A.: *Digital Image Processing.*, John Wiley and Sons, New York, 2014.

Beck, Conrad: *The Microscope: Theory and Practice.*, R. and J. Beck Ltd., London, United Kingdom, 2018.

Bederson, B. and H. Walther, *Advances in Atomic, Molecular, and Optical Physics,* Academic, New York, 2012.

Bergner, J., Gelbke: *Practical Photomicrography.*, Focal Press, New York, 2016.

Bloss, F. Donald: *An Introduction to the Methods of Optical Crystallography.*, Saunders College Publishing, Philadelphia, Pennsylvania, 2010.

Born, Max and Wolf, Emil: *Principles of Optics*, Cambridge University Press, Cambridge, United Kingdom, 2019.

Bracegirdle, Brian: *A History of Microtechnique*, Science Heritage Ltd., Lincolnwood, Illinois, 2017.

Bradbury, Savile: *The Evolution of the Microscope.*, Pergamon Press, Oxford, United Kingdom, 2017.

Brandreth, Gyles: *The Great Book of Optical Illusions.*, Sterling Publishing Company, New York, 2015.

Brau, C. A. : *Free-Electron Lasers,* Academic, New York, 2010.

Brown, T. G.: *The Optics Encyclopedia,* Wiley-VCH, Weinheim, 2014.

Buck, J.A.: *Fundamentals of Optical Fibers,* John Wiley and Sons, New York, 2015.

Cancellieri, G.: *Single-Mode Optical Fibres,* New York: Pergamon Press, 2011.

Chang, K. : *Handbook of Optical Components and Engineering,* Wiley, New York, 2013.

Cheo, P.K.: *Fiber Optics and Optoelectronics*, Prentice-Hall, Englewood Cliffs, New Jersey, 2015.

Cherry, Richard J.: *New Techniques of Optical Microscopy and Microspectroscopy*, CRC Press, Boca Raton, Florida, 2021.

Christian Hentschel: *Fiber Optics Handbook: An Introduction and Reference Guide to Fiber Optic Technology and Measurement Techniques*, West Germany: Hewlett Packard, 2018.

Conn, G. K. and Bradshaw F. J.: *Polarized Light in Metallography.*, Butterworths and Company Ltd., London, United Kingdom, 2012.

Dattolli, G. : *Lectures on the Free Electron Laser Theory and Related Topics,* World Scientific, London 2013.

Demtröder, W. : *Laserspektroscopie: Grundlagen und Techniken*, Springer, Berlin, 2017.

Desmarias, Louis: *Applied Electro Optics.*, Prentice Hall PTR, Upper Saddle River, New Jersey, 2018.

Diament, P.: *Wave Transmission and Fiber Optics*, Macmillan Publishing Company, New York, 2014.

Diels, J. C. and W. Rudolph, *Ultrashort Laser Pulse Phenomena,* Academic, New York, 2016.

Dougherty, E. R.: *Electronic Imaging Technology.*, SPIE Press, Bellingham, Washington, 2019.

Duarte, F. J. : *Tunable Laser Optics,* Elsevier Academic, New York, 2013.

Dunning, F. and R. G. Hulet : *Atomic, Molecular, and Optical Physics: Electromagnetic Radiation,* Academic, New York, 2017.

Ediger, M. : *Electro-Optics Handbook,* McGraw-Hill, New York, 2011.

Ekstrom, Michael P. : *Digital Image Processing Techniques*, Academic Press, Orlando, Florida, 2013.

Elion, G.R., and H.A. Elion: *Fiber Optics in Communications Systems,* Marcel Dekker, New York, 2018.

Fillard, J. P.: *Near Field Optics and Nanoscopy.*, World Scientific Publishing Company, Singapore, 2017.

Fowles, Grant R.: *Introduction of Modern Optics.*, Dover Publications, New York, 2015.

Gardener, Robert: *Experiments with Light and Mirrors.*, Enslow Publishers, Springfield, New Jersey, 2012.

Gerd Keiser, *Optical Fiber Communications*, New York, NY, USA: McGraw-Hill, 2016.

Ghatak, A.K., and K. Thyagarajan: *Contemporary Optics*, Plenum Press, New York, 2018.

Goldstein, D. J.: *Understanding the Light Microscope: A Computer-Aided Introduction.*, Academic Press, London, United Kingdom, 2019.

Gu, Min: *Advanced Optical Imaging Theory*, Springer-Verlag, Heidelberg, Germany, 2011.

Herman, Brian and Lemasters, John J. : *Optical Microscopy: Emerging Methods and Applications.*, Academic Press, San Diego, California, 2013.

Johnson, B. K.: *Optics and Optical Instruments: An Introduction with Special Reference to Practical Applications*, Dover Publications, New York, 2017.

Kao, C.K.: *Optical Fiber Technology*, McGraw-Hill, Inc., New York, 2011.

Keiser, G.E.: *Optical Fiber Communications*, McGraw-Hill, New York, 3rd ed., 2010.

Killen, H.B.: *Fiber Optic Communications*, Prentice-Hall, Englewood Cliffs, New Jersey, 2013.

Kingslake, Rudolf: *Lens Design Fundamentals.*, Academic Press, San Diego, California, 2018.

Kuehni, Rolf G.: *Color: An Introduction to Practice and Principles.*, John Wiley and Sons, New York, 2017.

Lakowicz, Joseph R.: *Principles of Fluorescence Spectroscopy*, Kluwer Academic/Plenum Publishers, New York, 2019.

Lin, C.: *Optoelectronic Technology and Lightwave Communications Systems*, Van Nostrand Reinhold, 2019.

Lloyd, Gill and Jefferis, David: *The History of Optics.*, Thomson Learning, New York, 2015.

Malies, Harold M.: *Applied Microscopy and Photomicrography.*, Fountain Press, London, United Kingdom, 2018.

Marcuse, D.: *Light Transmission Optics*, Van Nostrand Reinhold Company, New York, 2012.

Murphy, Douglas B.: *Fundamentals of Light Microscopy and Electronic Imaging.*, John Wiley and Sons, New York, 2011.

Nesse, William D.: *Introduction to Optical Mineralogy*, Oxford University Press, New York, 2013.

Ostrowsky, D.B.: *Fiber and Integrated Optics*, Plenum Press, New York, 2019.

Personick, S.D.: *Fiber Optics: Technology and Applications*, Plenum Press, New York, 2015.

Radziemski, L. J. ; R. W. Solarz: *Laser Spectroscopy and its Applications,* Marcel Dekker, New York, 2017.

Sudo, S.: *Optical Fiber Amplifiers: Materials, Devices, and Application Technologies*, Artech House, Norwood, Mass., 2016.

Tabib-Azar M.: *Integrated Optics, Microstructures, and Sensors*, Kluwer Academic Publisher, Boston, MA, 2013.

Ungar, S: *Fiber Optics Theory and Applications*, John Wiley and Sons, New York, 2011.

Waynant, R. and M. Ediger: *Electro-Optics Handbook,* McGraw-Hill, New York, 2014.

Index

M

O

P

Q

R

S

T

U

V

W

X